Geometry

Reproducibles

MP3497 Geometry
Author: Sara Freeman
Editor: Fran Lesser
Cover Design and Illustration: Cathy Tran
Interior Illustrations: Yoshi Miyake
Interior Design: Sara Freeman
Production: Linda Price, Sasha Govorova
Project Director: Linda C. Wood

ISBN: 978-0-7877-0594-7
Copyright © 2002
Milliken Publishing Company
a Lorenz company
P.O. Box 802
Dayton, OH 45401-0802
www.LorenzEducationalPress.com
Printed in the USA.
All rights reserved.

Developed for Milliken by The Woods Publishing Group, Inc.

The purchase of this book entitles the individual purchaser to reproduce copies by duplicating master or by any photocopy process for single classroom use.The reproduction of any part of this book for commercial resale or for use by an entire school or school system is strictly prohibited. Storage of any part of this book in any type of electronic retrieval system is prohibited unless purchaser receives written authorization from the publisher.

Milliken Publishing Company

Table of Contents

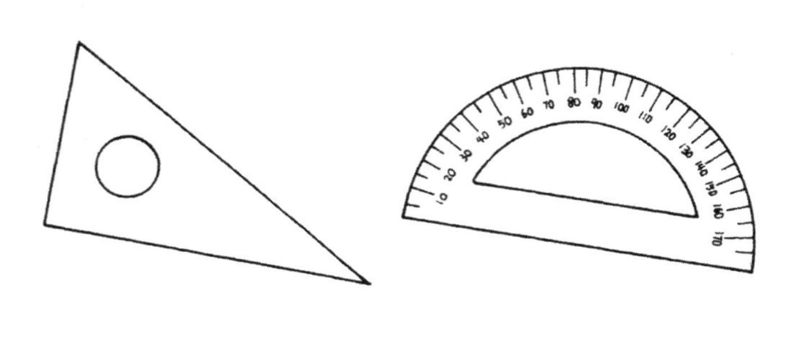

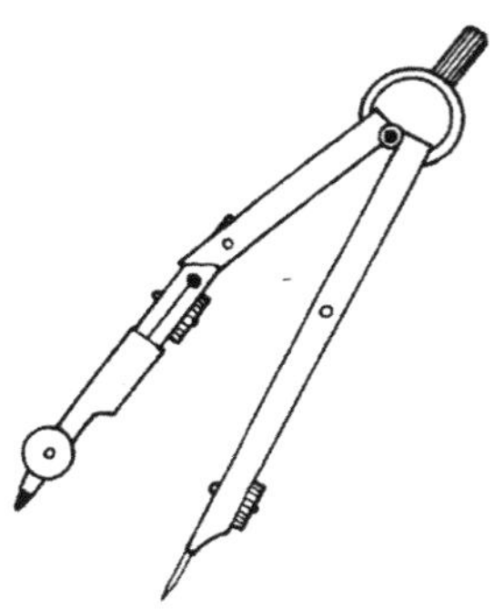

Name ______________________________

Geometry Basics

Remember

1. A *point* has position, but no dimension. • C **point C**
2. A *line* extends in one dimension. Points that lie on the same line are *collinear*. **$\overleftrightarrow{AB}$ or line k**
3. A *line segment* is part of a line with two endpoints. **$\overline{AB}$**
4. A *ray* is part of a line with one endpoint. **$\overrightarrow{AB}$**
5. An *angle* is formed by two rays or segments with the same endpoint. The endpoint is the *vertex*.

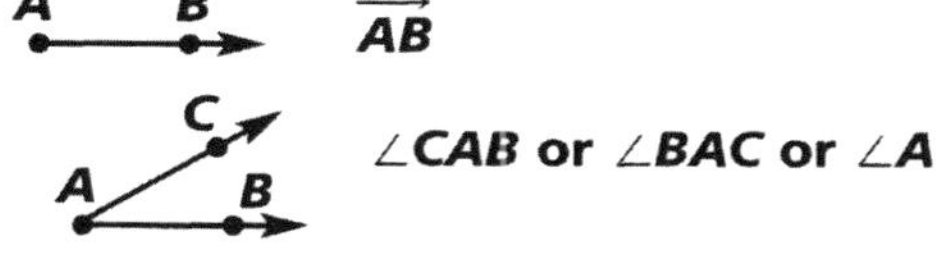

$\angle CAB$ or $\angle BAC$ or $\angle A$

6. A *plane* extends in two dimensions. Points that lie in the same plane are *coplanar*.

plane ABC or plane J

Refer to the diagrams and decide if each statement is true or false. If it is true, shade in the circle and write the letter on the puzzle blank. The puzzle answer is the name of a Greek mathematician and his books about geometry, number theory, and geometric algebra.

1. $\overleftrightarrow{EG}$ lies in plane R. (E)
2. $\overleftrightarrow{AB}$ lies in plane R. (B)
3. $\overleftrightarrow{MN}$ lies in plane S. (U)
4. D lies in plane R. (C)
5. H lies in plane R. (A)
6. C and E are coplanar. (L)
7. C and E are collinear. (Y)
8. M and O are collinear. (G)
9. M and L are collinear. (I)
10. $\angle CFE$ lies in plane R. (D)
11. $\angle JLM$ lies in plane S. (S)
12. Plane S intersects plane T at $\overline{JK}$. (E)
13. $\overleftrightarrow{AB}$ intersects plane R at F. (L)
14. L and P are in plane S. (P)
15. L and P are in plane T. (E)
16. $\overline{JK}$ is in plane S and plane T. (M)
17. $\overrightarrow{LN}$ is in plane S and plane T. (H)
18. L is the vertex of $\angle KLM$. (E)
19. $\angle BFD$ lies in plane R. (R)
20. $\overrightarrow{FC}$ and $\overrightarrow{FG}$ are sides of $\angle CFG$. (N)
21. $\angle DFG$ and $\angle GFD$ are the same angle. (T)
22. $\overline{LK}$ and $\overrightarrow{LN}$ are sides of $\angle KLN$. (S)

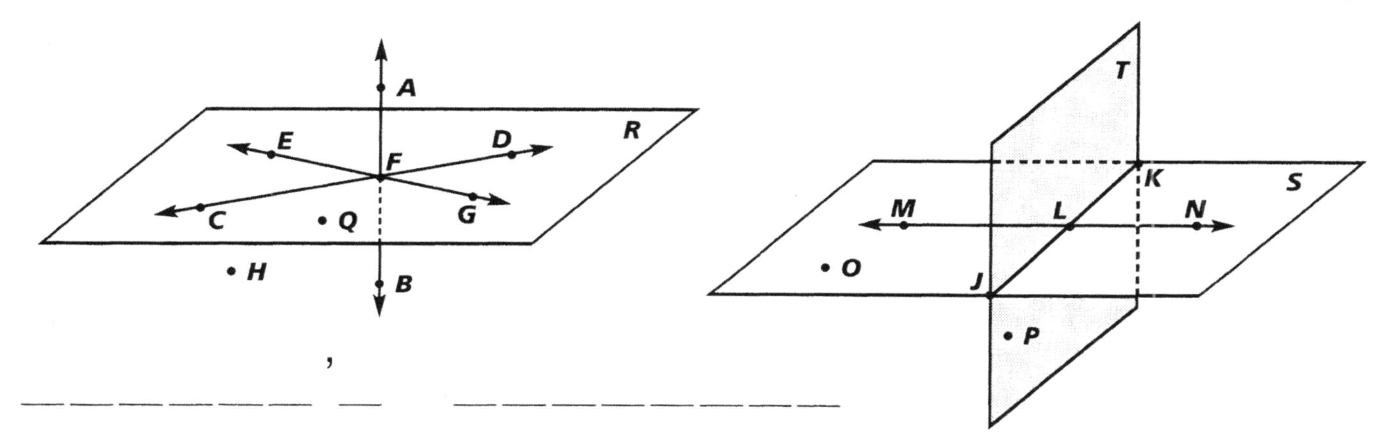

__ __ __ __ __ __ __ , __ __ __ __ __ __ __ __ __ __ __ __

© Milliken Publishing Company

Name ______________________________

Types of Angles

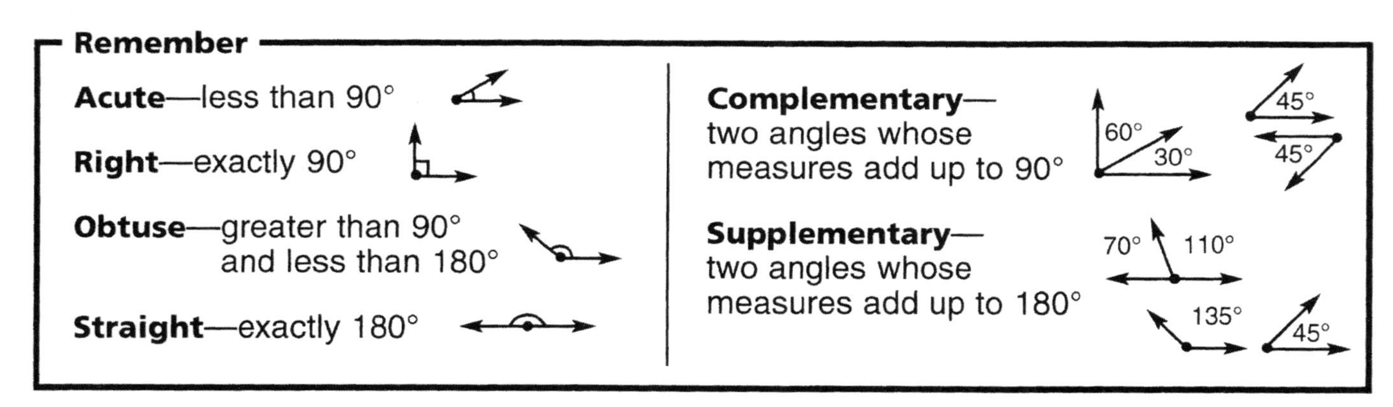

Remember

Acute—less than 90°

Right—exactly 90°

Obtuse—greater than 90° and less than 180°

Straight—exactly 180°

Complementary—two angles whose measures add up to 90°

Supplementary—two angles whose measures add up to 180°

Refer to the diagram and classify the angles. For 1–8, determine whether each angle is acute, right, obtuse, or straight. For 9–16, determine whether the two angles are complementary, supplementary, or neither. Circle the corresponding column letter and copy it onto the matching blanks below to complete the sentence.

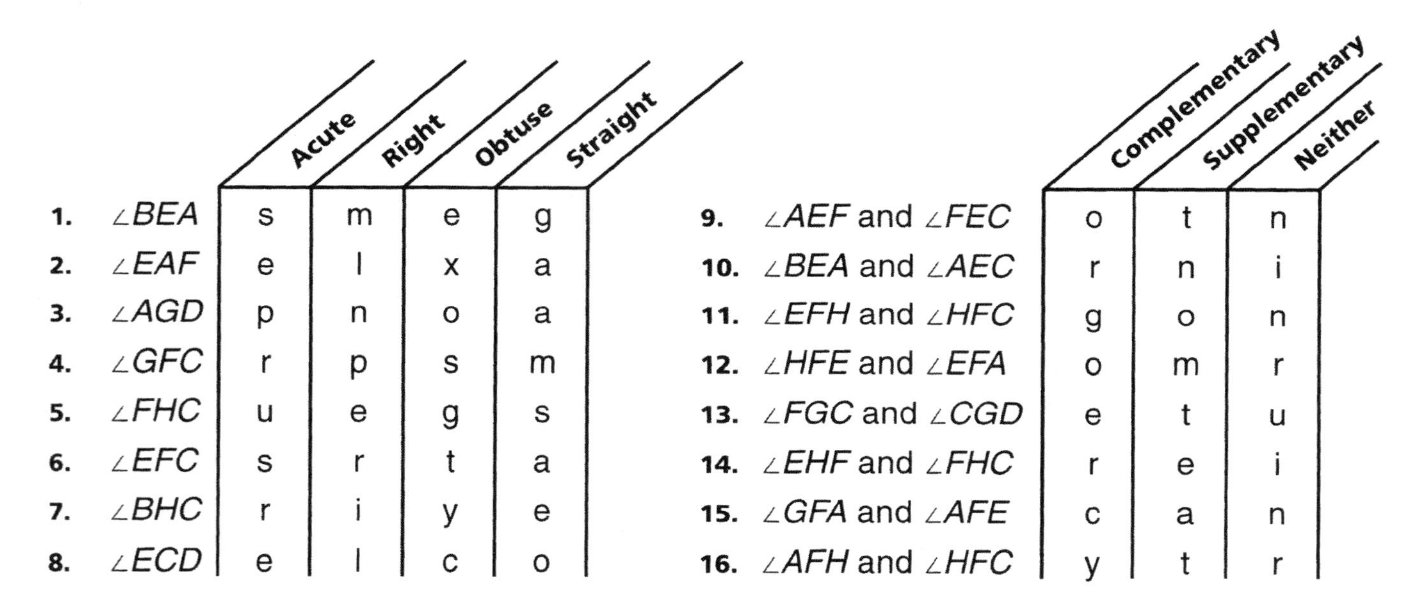

	Angle	Acute	Right	Obtuse	Straight
1.	∠BEA	s	m	e	g
2.	∠EAF	e	l	x	a
3.	∠AGD	p	n	o	a
4.	∠GFC	r	p	s	m
5.	∠FHC	u	e	g	s
6.	∠EFC	s	r	t	a
7.	∠BHC	r	i	y	e
8.	∠ECD	e	l	c	o

	Angles	Complementary	Supplementary	Neither
9.	∠AEF and ∠FEC	o	t	n
10.	∠BEA and ∠AEC	r	n	i
11.	∠EFH and ∠HFC	g	o	n
12.	∠HFE and ∠EFA	o	m	r
13.	∠FGC and ∠CGD	e	t	u
14.	∠EHF and ∠FHC	r	e	i
15.	∠GFA and ∠AFE	c	a	n
16.	∠AFH and ∠HFC	y	t	r

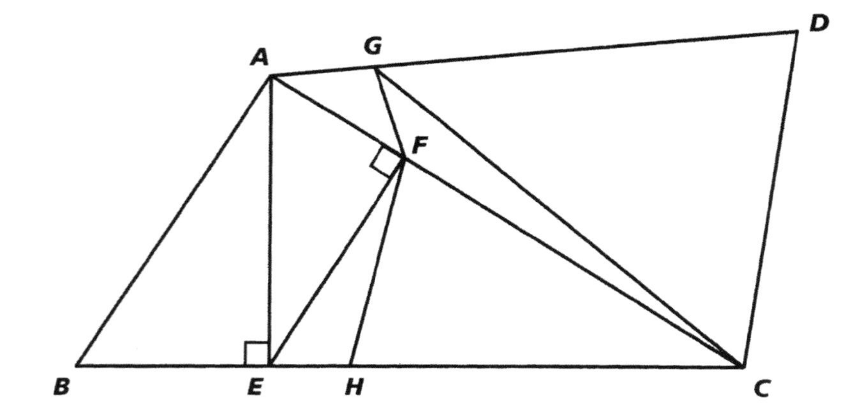

Angles that have the same

__ __ __ __ __ __ __
1 2 3 4 5 6 7

are called

__ __ __ __ __ __ __ __ __
8 9 10 11 12 13 14 15 16

angles.

© Milliken Publishing Company

Name ______________________________

Algebra Angle Measures

Example

The supplement of an angle is 30° less than twice the measure of the angle itself. Find the angle.

1. Make a sketch, using *x* to represent the angle. (Complementary angles add up to 90°; supplementary angles add up to 180°.)

$2x - 30$ / x

2. Write an equation. $x + 2x - 30 = 180$

3. Solve for *x*.

$$x + 2x - 30 = 180$$
$$3x - 30 = 180$$
$$3x = 210$$
$$x = 70°$$

4. Check your answer.

The measure of the angle is 70°.
The supplement is (2 x 70) – 30 = 110°.
70° + 110° = 180°

Read each problem and draw a line to its matching sketch. Write an equation for the problem, using x for the angle. Solve for x. When you finish, find and circle your answer in the box below.

1. The supplement of an angle is twice the measure of the angle itself. Find the angle.

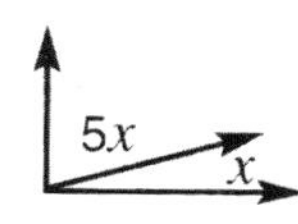

2. The complement of an angle is five times the measure of the angle itself. Find the angle.

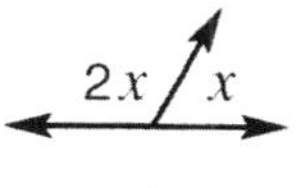

3. The complement of an angle is 10° less than the measure of the angle itself. Find the angle.

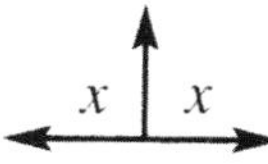

4. The supplement of an angle is 20° more than the measure of the angle itself. Find the angle.

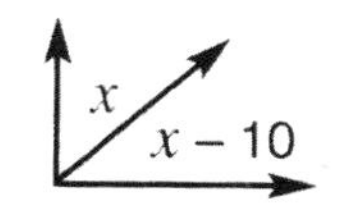

5. Two angles are congruent and complementary. Find their measures.

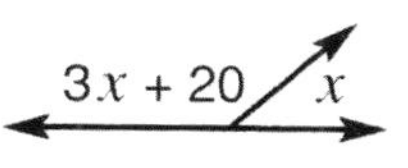

6. Two angles are congruent and supplementary. Find their measures.

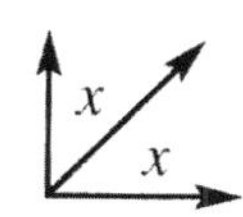

7. The supplement of an angle is 20° more than three times the measure of the angle itself. Find the angle.

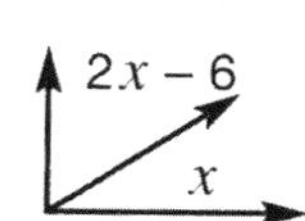

8. The complement of an angle is 6° less than twice the measure of the angle itself. Find the angle.

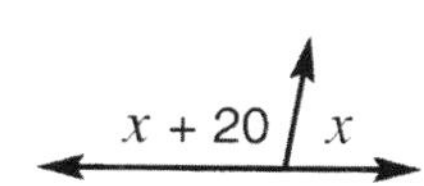

15°	32°	40°	45°	50°	60°	70°	80°	90°

© Milliken Publishing Company

Name ______________________________

Angles Formed by Parallel, Perpendicular, and Intersecting Lines

Remember

1. If two parallel lines are cut by a transversal, the resulting angles are either congruent or supplementary.

Congruent angles:	**Supplementary angles:**
Vertical angles ($\angle 2 \cong \angle 3$)	**Adjacent angles** ($m\angle 1 + m\angle 3 = 180°$)
Corresponding angles ($\angle 1 \cong \angle 5$)	**Same side interior angles** ($m\angle 3 + m\angle 5 = 180°$)
Alternate interior angles ($\angle 4 \cong \angle 5$)	**Same side exterior angles** ($m\angle 2 + m\angle 8 = 180°$)
Alternate exterior angles ($\angle 1 \cong \angle 8$)	

2. If the transversal is perpendicular, the angles formed are right angles. ($m\angle 9 = 90°$)

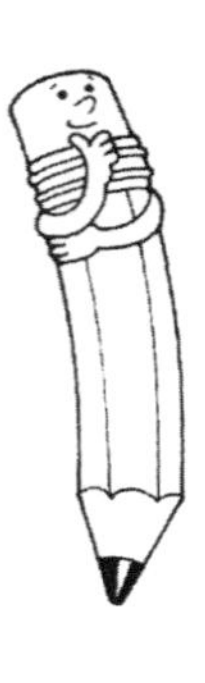

Refer to the diagram to complete each sentence. Fill in the missing angle measure and type of angle. When you finish, circle each answer in the box below.

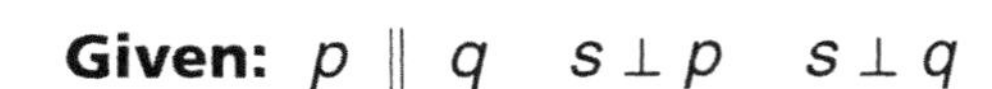

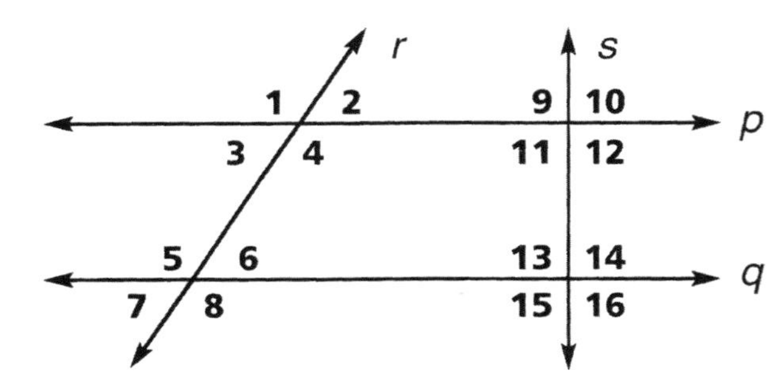

1. If $m\angle 1 = 120°$, $m\angle 2 =$ ______ because they are ______________ angles.
2. If $m\angle 5 = 130°$, $m\angle 8 =$ ______ because they are ______________ angles.
3. If $m\angle 4 = 125°$, $m\angle 6 =$ ______ because they are same side ______________ angles.
4. If $m\angle 4 = 125°$, $m\angle 8 =$ ______ because they are ______________ angles.
5. If $m\angle 2 = 45°$, $m\angle 7 =$ ______ because they are alternate ______________ angles.
6. If $m\angle 3 = 50°$, $m\angle 6 =$ ______ because they are ______________ interior angles.
7. If $m\angle 7 = 42°$, $m\angle 1 =$ ______ because they are ______________ exterior angles.
8. The measures of $\angle 9$ through $\angle 16 =$ ______ because they are all __________ angles.

45°	50°	55°	60°	90°	125°	130°	138°
adjacent	alternate	corresponding	exterior	interior	right	same side	vertical

© Milliken Publishing Company

Name ______________________________

Angle and Segment Bisectors

Remember

1. An **angle bisector** is a ray that divides an angle into two adjacent and congruent angles.

Examples:

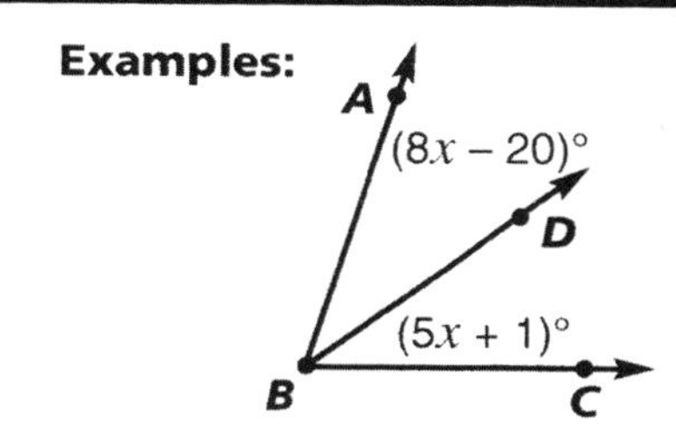

$\overrightarrow{BD}$ **bisects** $\angle ABC$. **Therefore,**

$$8x - 20 = 5x + 1$$
$$3x = 21$$
$$x = 7$$

$m\angle ABD = 8(7) - 20 = 36°$
$m\angle DBC = 5(7) + 1 = 36°$
$m\angle ABC = 36° + 36° = 72°$

2. A **segment bisector** is a segment, ray, line, or plane that divides a segment into two congruent segments.

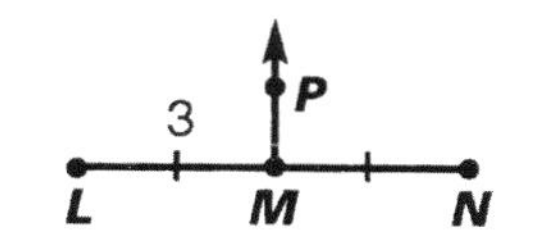

$\overrightarrow{MP}$ **bisects** $\overline{LN}$. **Given:** $LM = 3$
Thus: $MN = 3$ $LN = 6$

Find the missing angle measures or segment lengths. Shade in your answers in the box to find the term for the point that divides a segment into two congruent segments.

1. $\overrightarrow{BD}$ bisects $\angle ABC$.
$m\angle ABD$ = ______
$m\angle ABC$ = ______

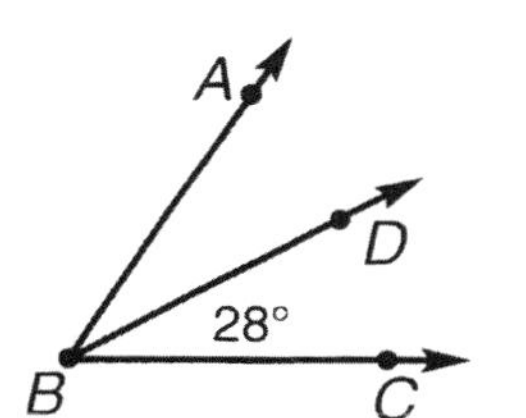

2. $\overrightarrow{FH}$ bisects $\angle EFG$.
$m\angle EFH$ = ______
$m\angle HFG$ = ______

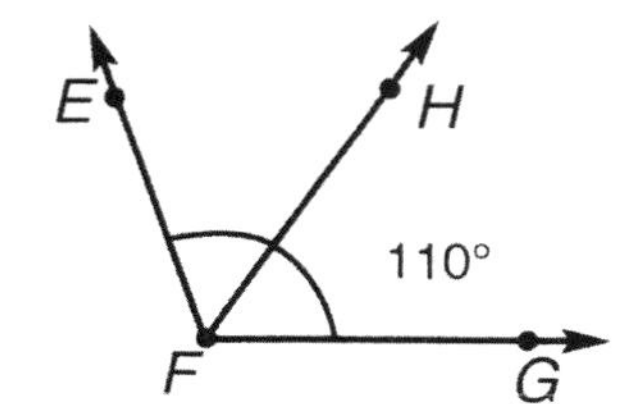

3. $\overrightarrow{JL}$ bisects $\angle IJK$.
$m\angle IJL$ = ______
$m\angle IJK$ = ______

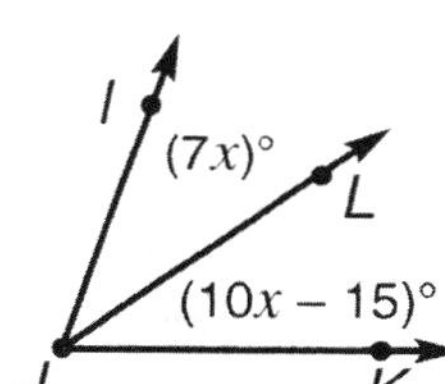

4. $\overrightarrow{NP}$ bisects $\angle MNO$.
$m\angle PNO$ = ______
$m\angle MNO$ = ______

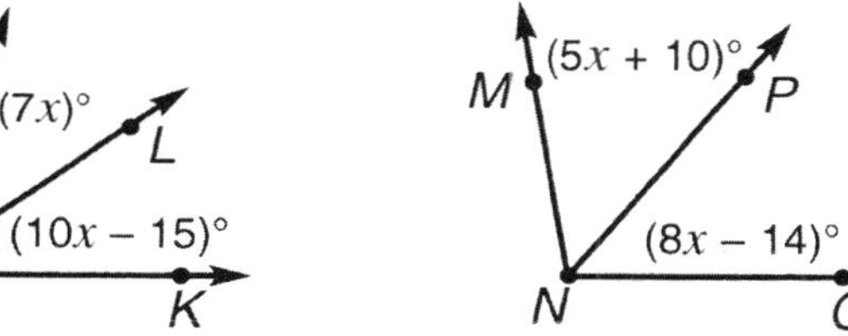

5. $\overrightarrow{RT}$ bisects $\angle QRS$.
$m\angle TRS$ = ______
$m\angle QRS$ = ______

Q $(14x)°$ T
$(11x + 9)°$
R S

6. $\overrightarrow{CG}$ bisects $\overline{AB}$. $\overrightarrow{EF}$ bisects $\overline{DC}$.
$CB = 11$ $DE = 4$ EC = ____
AB = ______ AD = ____

F G
4 11
A D E C B

7. $\overleftrightarrow{YZ}$ bisects $\overline{UV}$. $\overleftrightarrow{WX}$ bisects $\overline{UY}$.
UW = ______ YV = ______
UV = ______

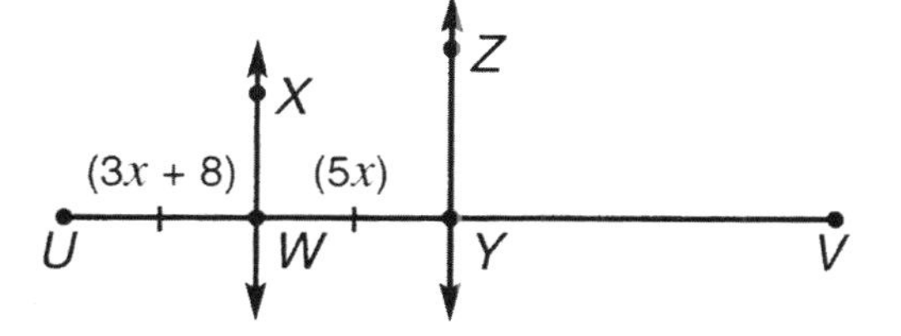

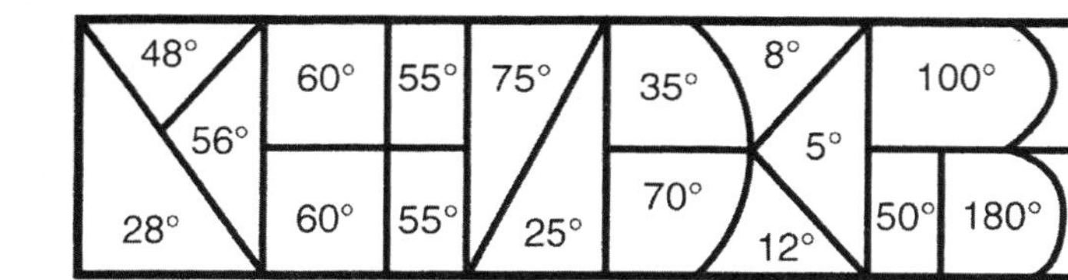

© Milliken Publishing Company

Name ______________________________

Angles of a Triangle

Remember

1. The three interior angles of a triangle add up to 180°. ($m\angle 1 + m\angle 2 + m\angle 3 = 180°$)
2. If each side is extended in one direction, exterior angles are formed. ($\angle 4$, $\angle 5$, $\angle 6$)
3. An interior angle and its adjacent exterior angle are supplements. ($m\angle 1 + m\angle 4 = 180°$)

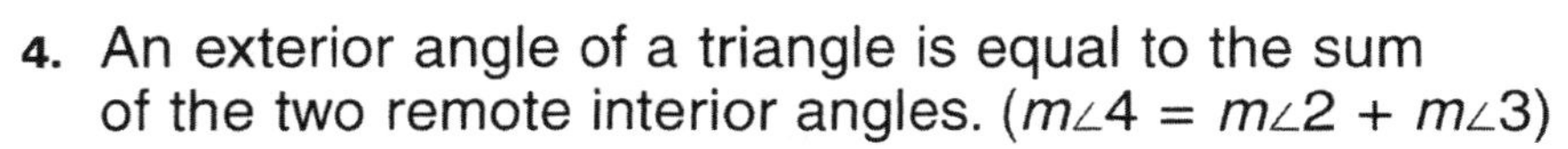

4. An exterior angle of a triangle is equal to the sum of the two remote interior angles. ($m\angle 4 = m\angle 2 + m\angle 3$)

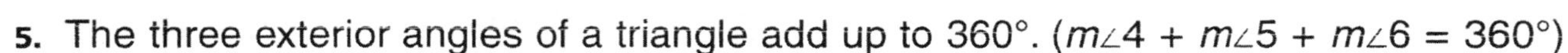

5. The three exterior angles of a triangle add up to 360°. ($m\angle 4 + m\angle 5 + m\angle 6 = 360°$)

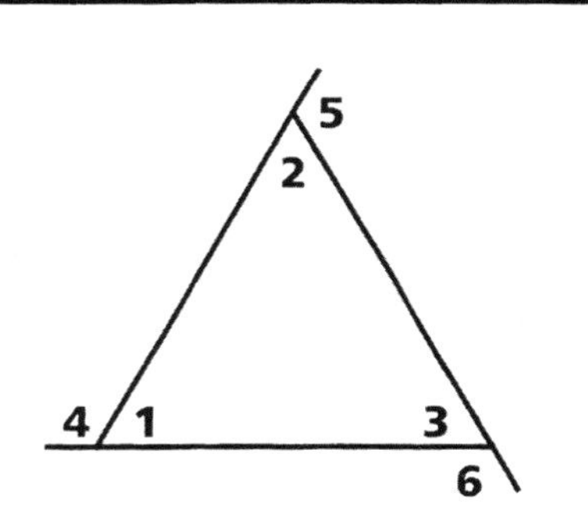

Find the missing angle measures. Follow your answers in alphabetical order through the maze.

1.

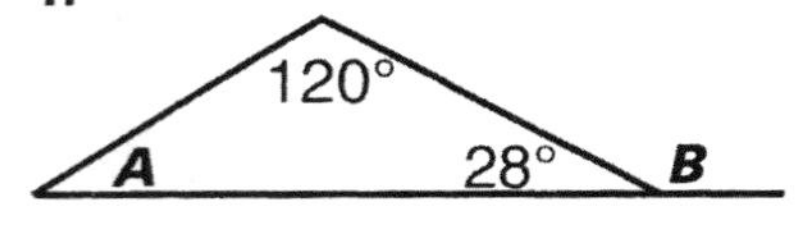

2.

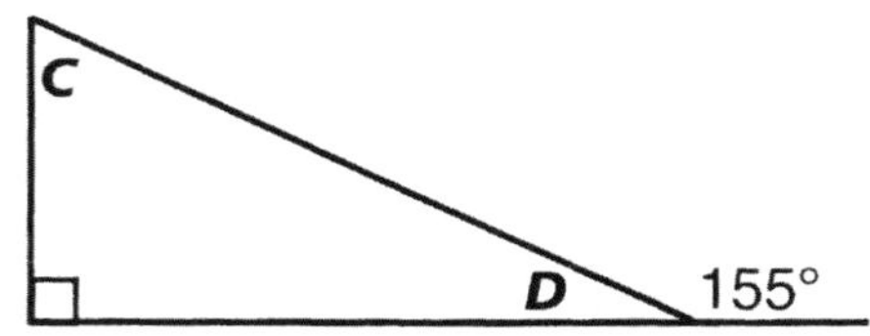

3.

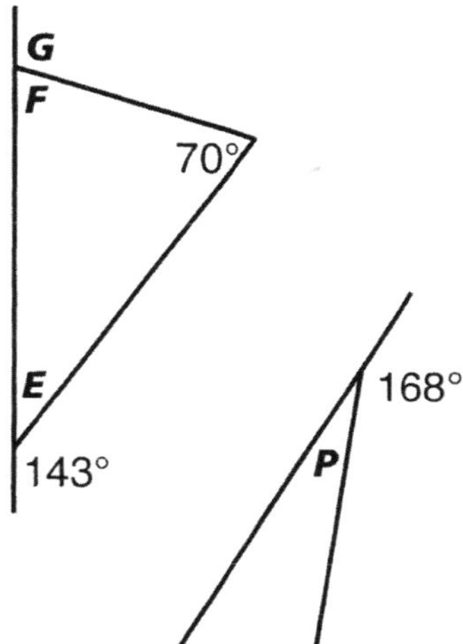

4.

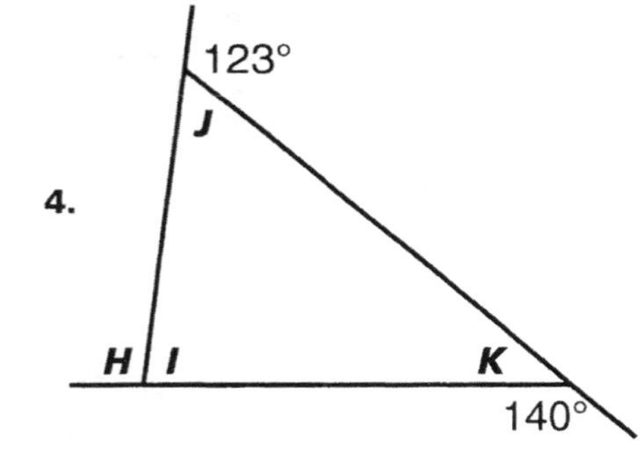

5.

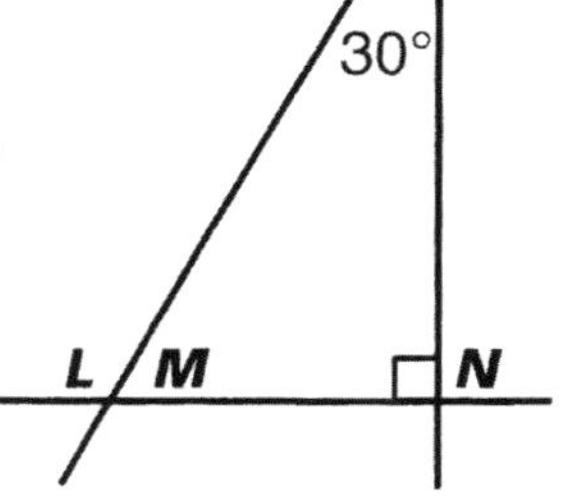

6.

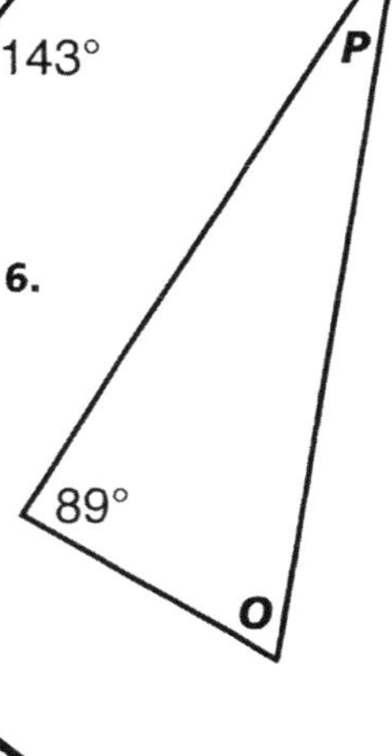

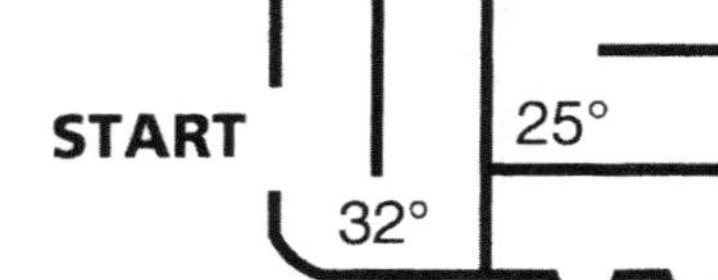

Name ______________________________

Triangle Inequalities

Remember

1. In a triangle, the longest side is opposite the largest angle and the shortest side is opposite the smallest angle.

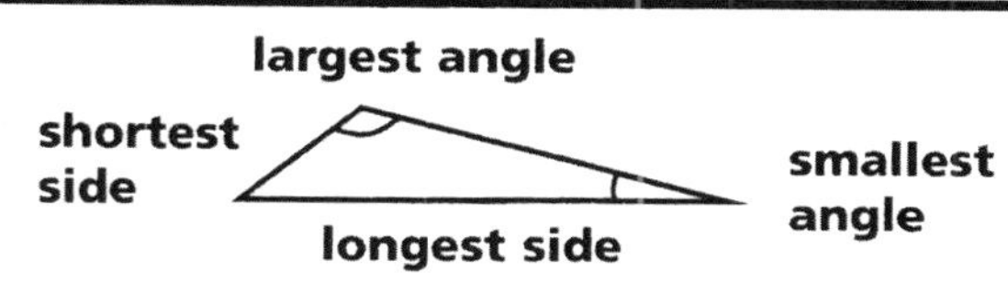

2. When given the lengths of two sides of a triangle, the length of the third side must be greater than their difference, but less than their sum.

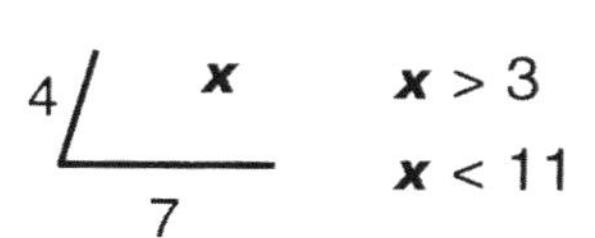

$x > 3$

$x < 11$

A. Circle the letter of the smallest angle or shortest side of each triangle. The diagrams may not be drawn to scale; base your answers on the measurements given.

1.

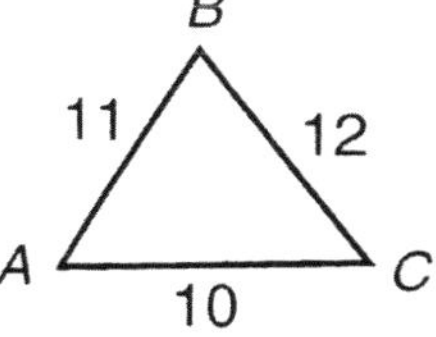

2.

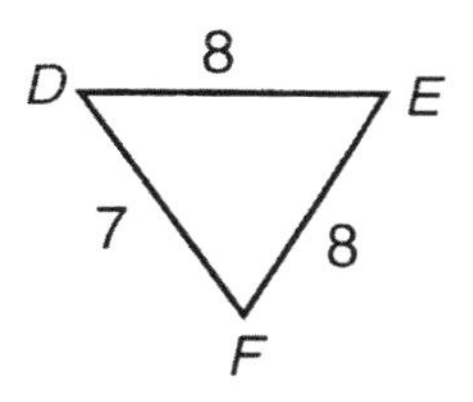

3.

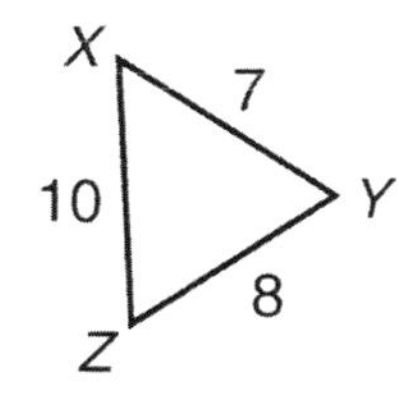

4.

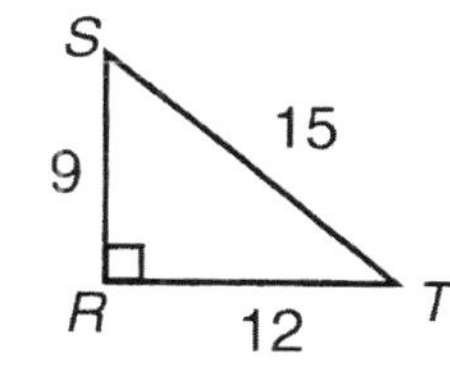

5.

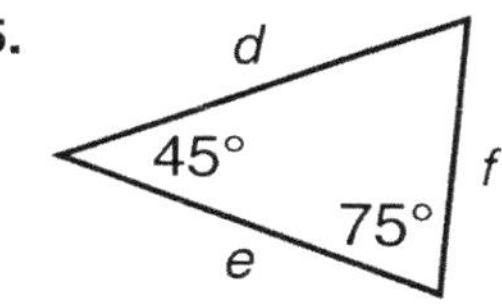

6.

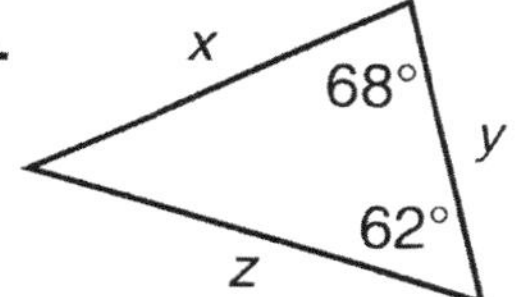

7.

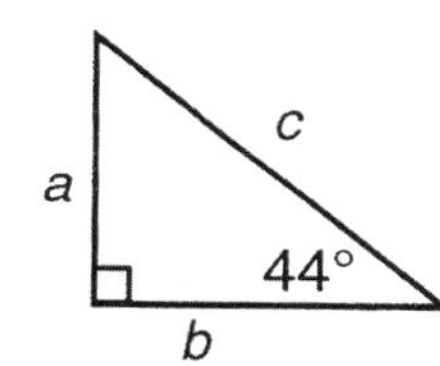

8.

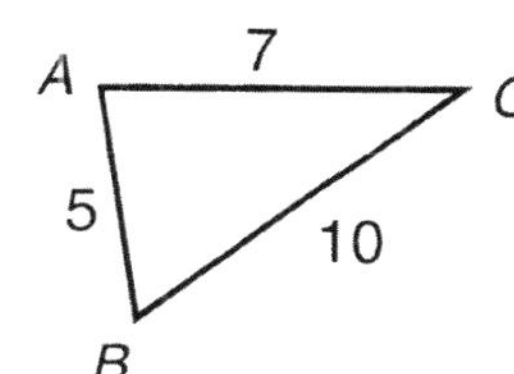

9.

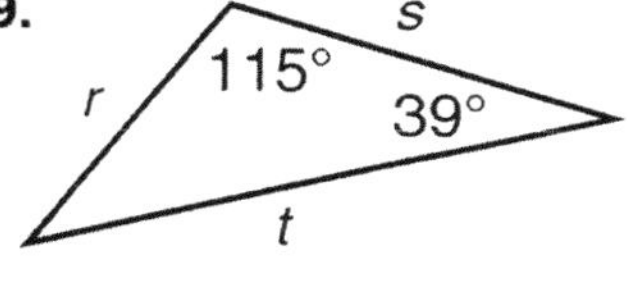

10.

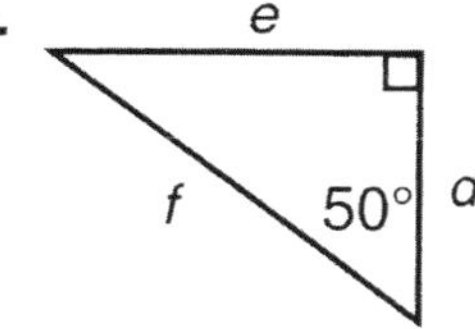

B. Complete the chart.

	Lengths of two sides of a triangle	Length of third side must be greater than	less than
11.	8 and 12		
12.	15 and 17		
13.	20 and 25		
14.	3 and 4		
15.	9 and 9		

C. Create a design by drawing straight lines to connect your answers in the order of the problems. Begin at the star.

B ★
Z •
f •
a •
s •
4 •
2 •
5 •
1 •
0 •

• E • T • y • C • d • 20 • 32 • 45 • 7 • 18

© Milliken Publishing Company

Name ______________________________

Types of Triangles

Remember

1. An **equiangular** triangle has three congruent angles.

An **acute** triangle has three acute angles.

A **right** triangle has one right angle.

An **obtuse** triangle has one obtuse angle.

2. An **equilateral** triangle has three congruent sides.

An **isosceles** triangle has at least two congruent sides.

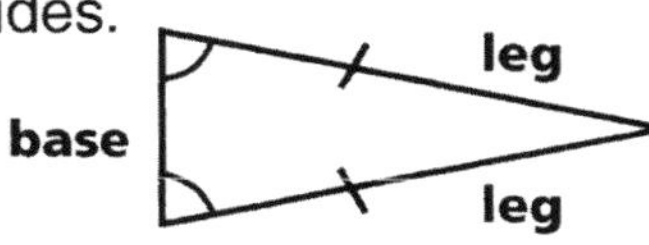

A **scalene** triangle has no congruent sides.

If an **isosceles** triangle has exactly two congruent sides, those sides are the legs and the third side is the base. The two base angles are congruent.

Solve for the missing triangle side lengths or angle measures. Classify each triangle and shade the corresponding column letters in the chart. There will be two or four chart answers per triangle.

1.

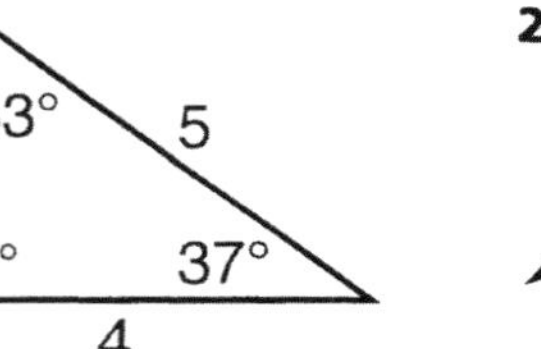

2. 120°, 8, s

3.

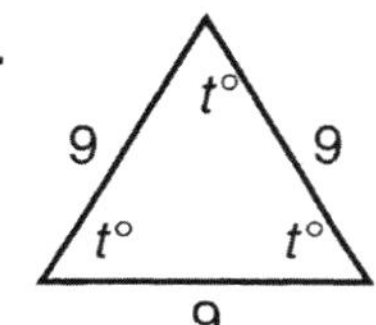

4.

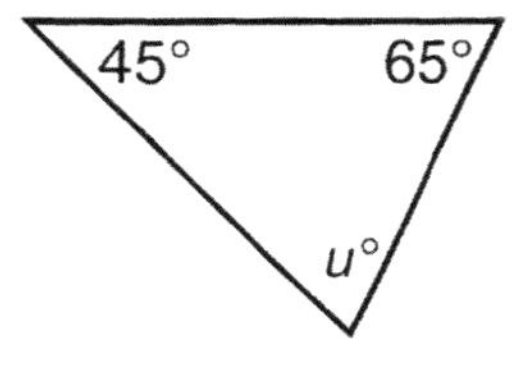

5. 30°, v°, v°

6.

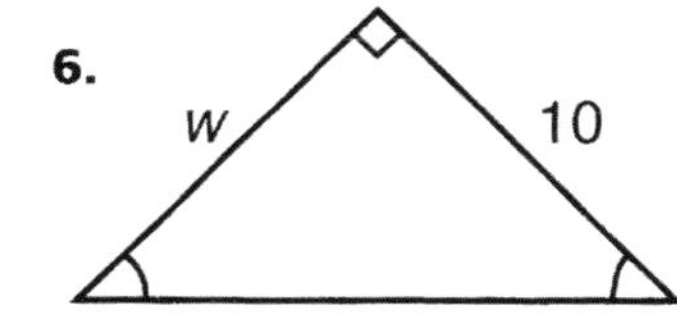

7. x°, 50°, 20°

8.

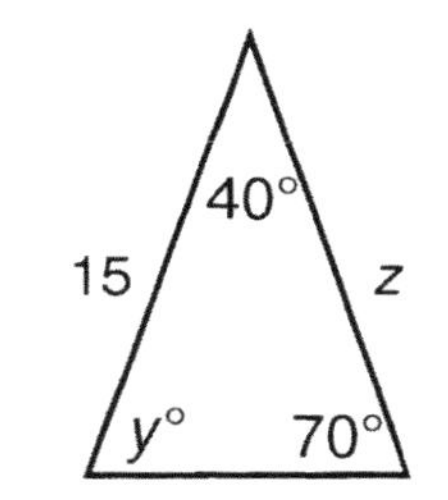

	equiangular	acute	right	obtuse	equilateral	isosceles	scalene
1.	T	R	P	A	N	G	Y
2.	B	A	L	T	E	H	R
3.	A	G	Z	T	O	R	M
4.	F	E	I	C	S	T	A
5.	J	N	K	B	O	T	E
6.	Q	R	H	A	U	E	L
7.	I	D	V	O	N	G	R
8.	W	E	X	T	R	M	Y

The formula $c^2 = a^2 + b^2$ tells the relationship between the lengths of the sides in a right triangle. What is it known as? To find out, write the shaded chart letters in order.

__ __ __ __ __ __ __ __ __ __ __

__ __ __ __ __ __ __

© Milliken Publishing Company

Name ______________________________

Cartesian Coordinates

Remember

1. The first number in an ordered pair is the x-coordinate. It tells where a point is located along the horizontal axis.
2. The second number in an ordered pair is the y-coordinate. It tells where a point is located along the vertical axis.

Example: △***ABC***

A (−5, 3) ***B*** (−5, 1) ***C*** (−1, 1)

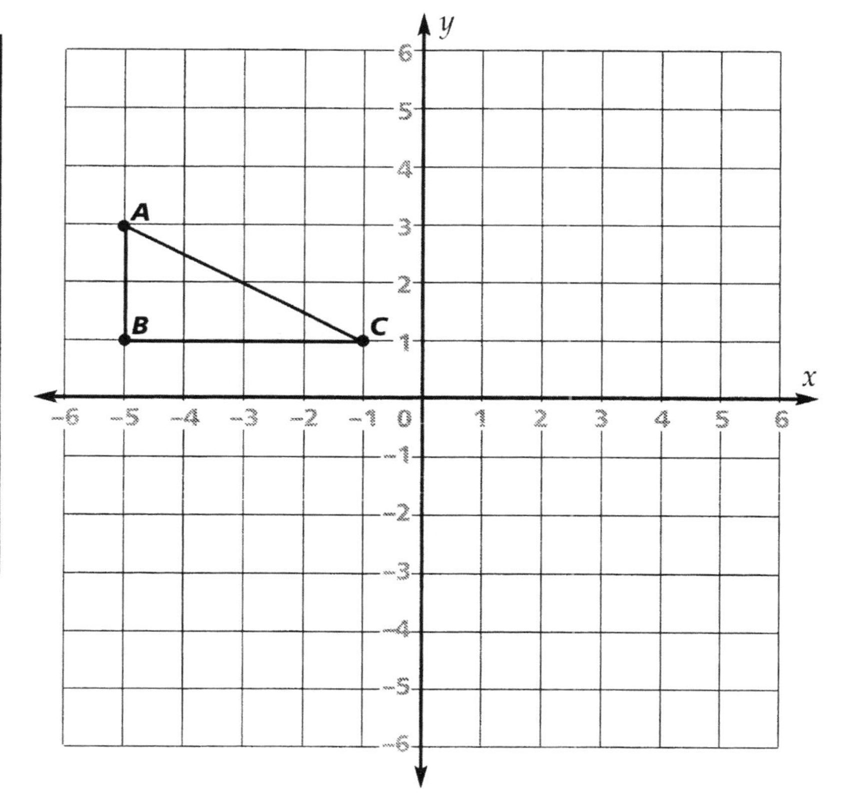

Plot, label, and connect these points to create four triangles.

1. △***DEF*** ***D*** (2, −2) ***E*** (2, −4) ***F*** (6, −4)

2. △***GHI*** ***G*** (−3, −5) ***H*** (−1, −5) ***I*** (−1, −1)

3. △***JKL*** ***J*** (5, 3) ***K*** (5, 1) ***L*** (1, 1)

4. △***MNO*** ***M*** (1, 6) ***N*** (1, 5) ***O*** (3, 5)

Compare the position and size of each new triangle on the graph to that of △*ABC*. Draw lines to match each triangle below to the description of its transformation of △*ABC*. The uncrossed letters will spell out a message.

5. △***DEF*** • **S** **U** **M** • a reflection (flip) of △***ABC*** over the y-axis

6. △***GHI*** • **E** **P** • a rotation (turn) of △***ABC***

7. △***JKL*** • **E** **D** • a dilation (scaled enlargement or reduction) of △***ABC***

8. △***MNO*** • **I** **R** • a translation (slide) of △***ABC***

___ ___ ___ ___ ___ !

© Milliken Publishing Company

Name ______________________________

The Midpoint Formula

Remember

To find the midpoint between two ordered pairs, add the x-coordinates and divide by 2, then add the y-coordinates and divide by 2.

Midpoint $= \left(\frac{x_1 + x_2}{2}, \frac{y_1 + y_2}{2} \right)$

Example: Find the midpoint between ($^-5$, $^-3$) and (0, 7).

$M = \left(\frac{^-5 + 0}{2}, \frac{^-3 + 7}{2} \right) = \left(\frac{^-5}{2}, \frac{4}{2} \right) = (^-2.5, 2)$

Find the midpoints for these sets of ordered pairs. Then graph each segment formed by the two ordered pairs, checking that the midpoint divides the segment into two congruent segments. You will reveal a mathematical symbol that was introduced in 1525.

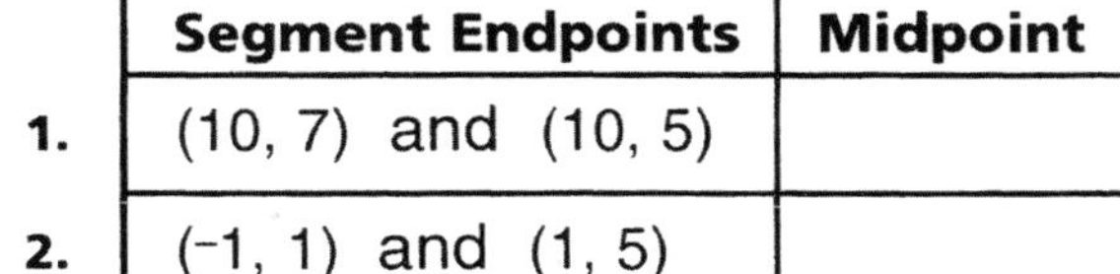

	Segment Endpoints	Midpoint
1.	(10, 7) and (10, 5)	
2.	($^-1$, 1) and (1, 5)	
3.	($^-5$, $^-7$) and ($^-8$, $^-1$)	
4.	(0, 7) and (10, 7)	
5.	($^-1$, 1) and ($^-5$, $^-7$)	
6.	($^-7$, 1) and ($^-5$, $^-3$)	
7.	(1, 5) and (10, 5)	

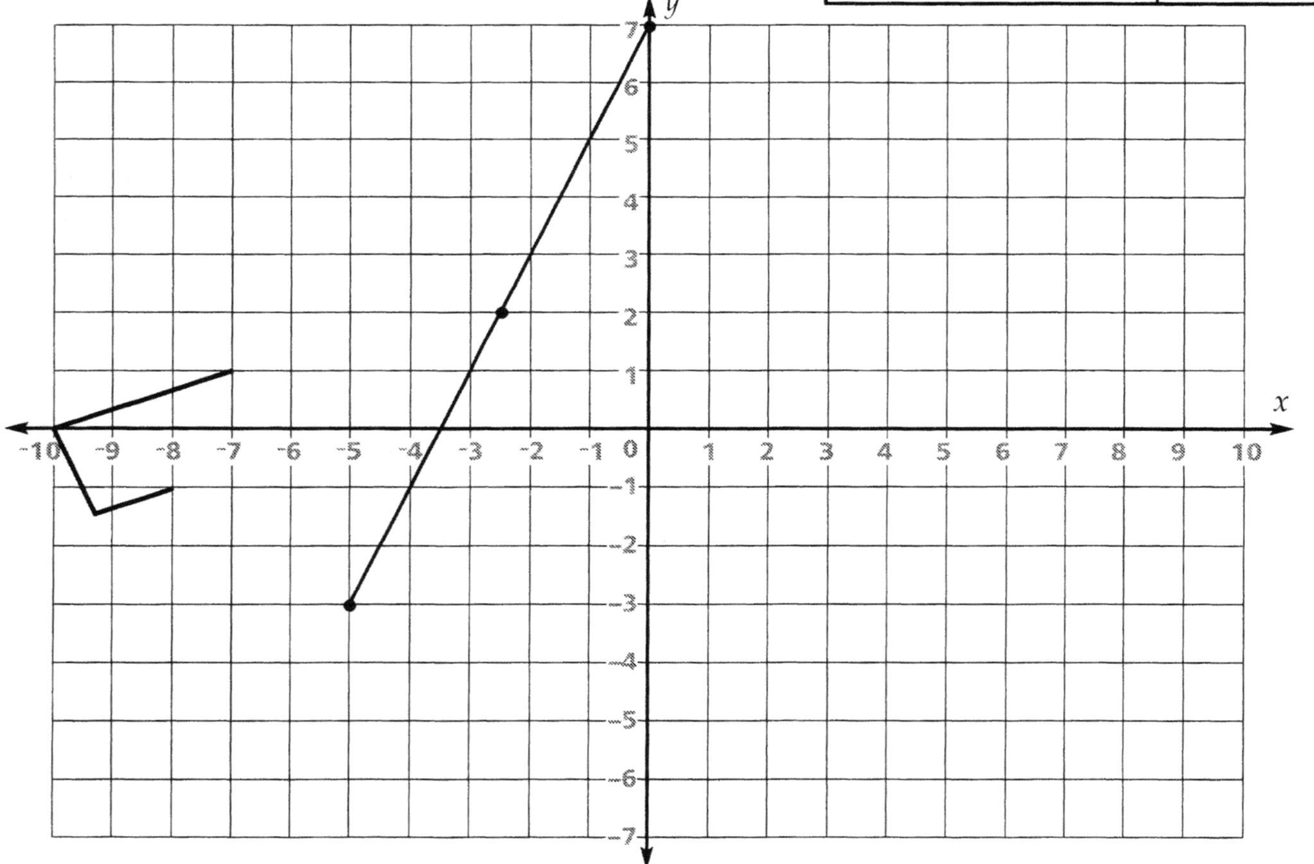

Name ______________________________

Special Segments in Triangles

Remember

1. A **median** of a triangle is a segment from a vertex to the midpoint of the opposite side.

X Y M Z

2. An **altitude** of a triangle is a perpendicular segment from a vertex to the opposite side (or an extension of it).

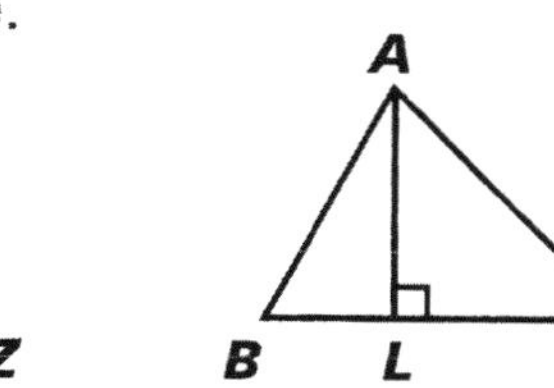

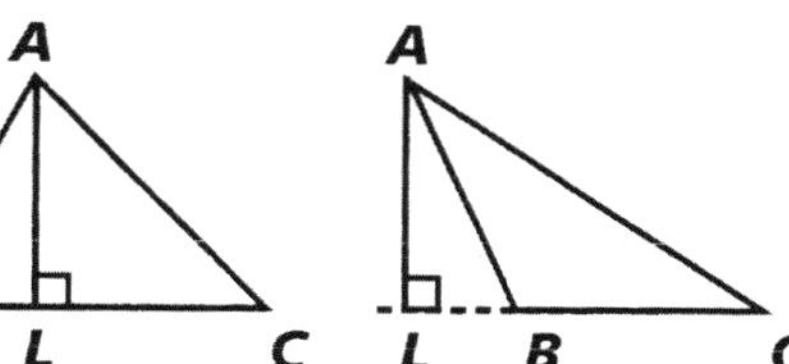

3. A **midline** of a triangle connects two midpoints of two sides and is parallel to the third side. Its length is half the length of the third side.

$\overline{PQ} \parallel \overline{JK}$

$PQ = \frac{1}{2}JK$

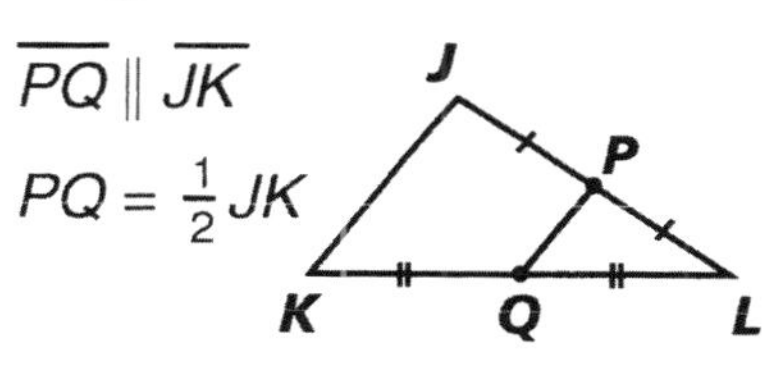

Refer to the diagrams to complete these sentences. Find the answers in the decoder and use them to reveal the civilization that studied triangles about 3,700 years ago.

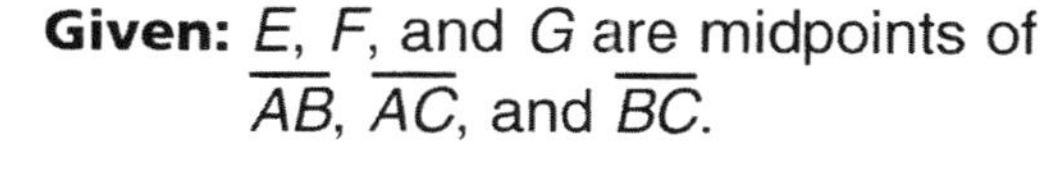

Given: *E*, *F*, and *G* are midpoints of $\overline{AB}$, $\overline{AC}$, and $\overline{BC}$.

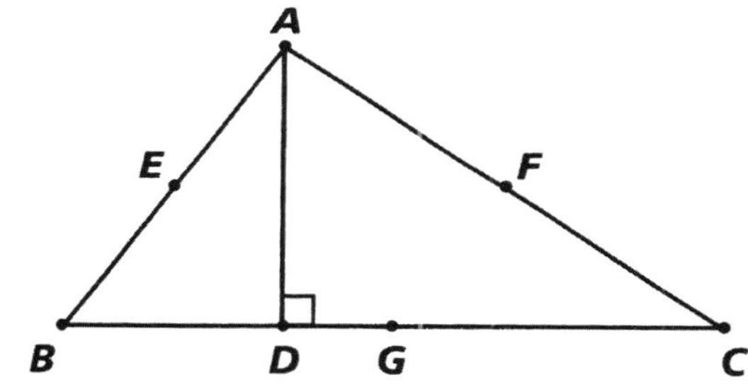

1. $\overline{AD}$ is an ______________________.
2. $\overline{BF}$ is a ______________________.
3. $\overline{EG}$ is a ______________________.
4. The median from C is segment _______.
5. If $EF = 10$, then segment _______ = 20.
6. If $AC = 16$, then the length of $\overline{FC}$ is ______.
7. If $AB = 12$, then the length of $\overline{FG}$ is ______.

Given: *M*, *N*, and *O* are midpoints of $\overline{JK}$, $\overline{JL}$, and $\overline{KL}$.

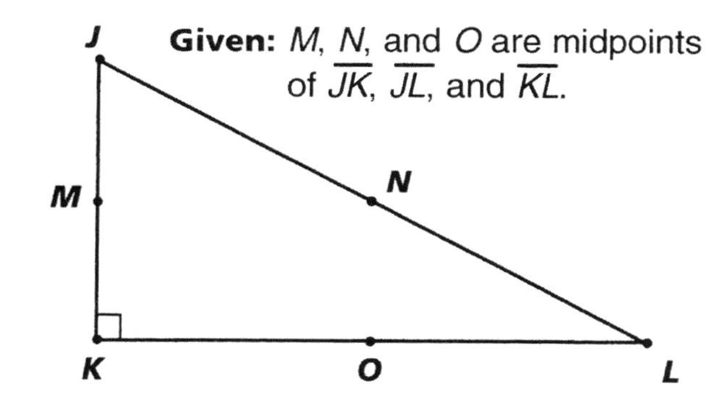

8. The altitude from J is segment _______.
9. If $JM = 10$, then the length of $\overline{MK}$ is ______.
10. If $KL = 30$, then the length of $\overline{MN}$ is ______.
11. *MO* is half the length of segment _______.
12. If the perimeter of $\triangle JKL$ were 60 units, the perimeter of $\triangle MNO$ would be _______ units.

6	8	10	15	20	30	32	*BC*	*CE*	*CG*	*JK*	*JL*	*JO*	median	altitude	midline
a	b	B	e	f	h	H	i	l	m	n	o	P	s	T	y

___ ___ ___   ___ ___ ___ ___ ___ ___ ___ ___ ___ ___ ___

1 12 10   9 7 6 3 4 11 8 5 7 8 2

© Milliken Publishing Company

Name ______________________________

Congruent Triangles— SSS, SAS, ASA

Remember

Two figures are *congruent* if they are the same shape and size. The two figures have corresponding sides and corresponding angles that are congruent.

Side-Side-Side (SSS) Congruence—If three sides of one triangle are congruent to three sides of another triangle, then the triangles are congruent.

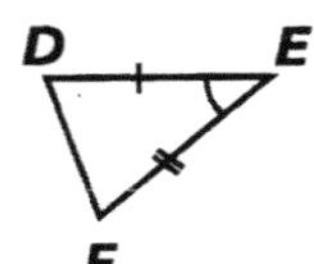

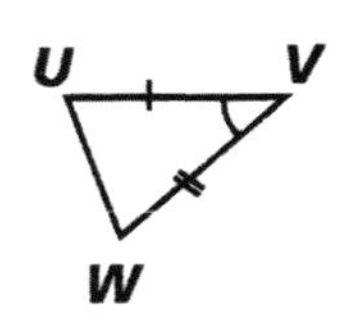

Side-Angle-Side (SAS) Congruence—If two sides and the included angle of one triangle are congruent to two sides and the included angle of another triangle, then the triangles are congruent.

Z Y G X H I

Angle-Side-Angle (ASA) Congruence—If two angles and the included side of one triangle are congruent to two angles and the included side of another triangle, then the triangles are congruent.

Determine which method if any can prove the triangles are congruent. Shade in the matching column letters and copy them onto the blanks to reveal a message.

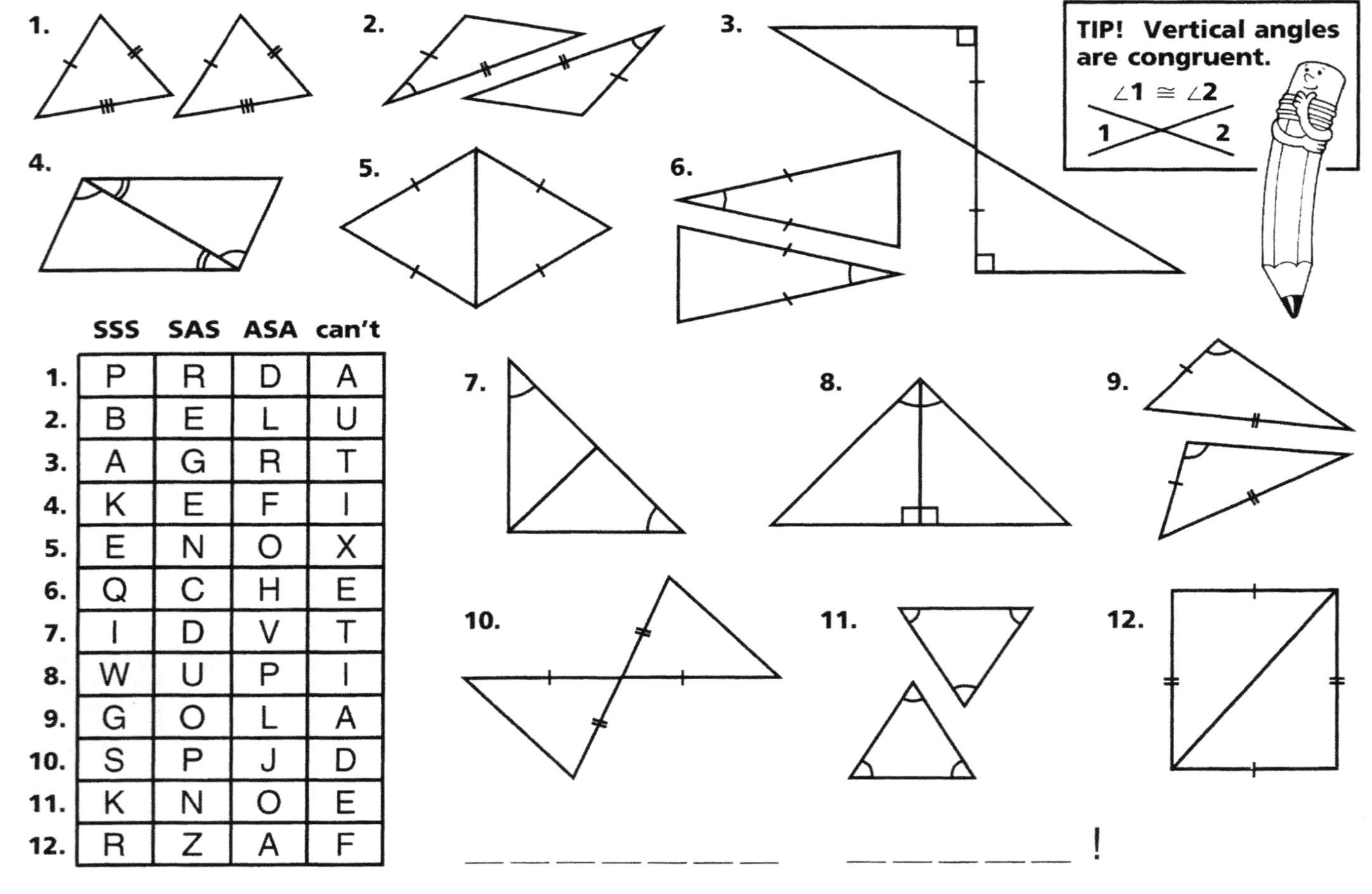

	SSS	SAS	ASA	can't
1.	P	R	D	A
2.	B	E	L	U
3.	A	G	R	T
4.	K	E	F	I
5.	E	N	O	X
6.	Q	C	H	E
7.	I	D	V	T
8.	W	U	P	I
9.	G	O	L	A
10.	S	P	J	D
11.	K	N	O	E
12.	R	Z	A	F

_ !

© Milliken Publishing Company

Name ______________________________

Congruent Triangles—AAS, HL

Remember

Angle-Angle-Side (AAS) Congruence—If two angles and a non-included side of one triangle are congruent to two angles and a non-included side of another triangle, then the two triangles are congruent.

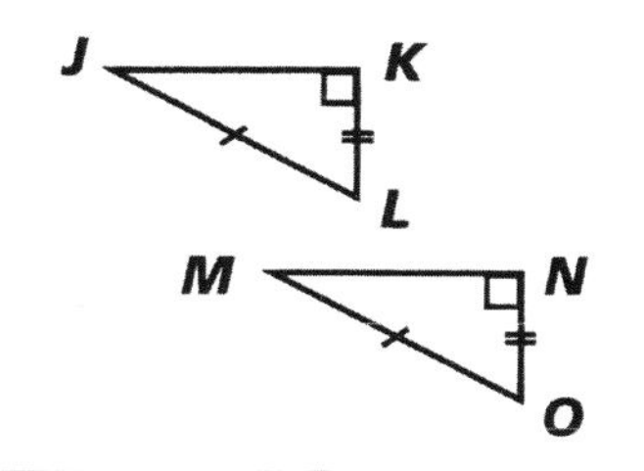

Hypotenuse-Leg (HL) Congruence—If the hypotenuse and a leg of one right triangle are congruent to the hypotenuse and a leg of another right triangle, then the two triangles are congruent.

In a right triangle, the sides that form the right angle are *legs*. The side opposite the right angle is the *hypotenuse*.

Determine which methods if any can prove the triangles are congruent. There may be more than one answer. Shade in the matching column letters. Copy the letters onto the blanks to reveal the riddle answer.

1.

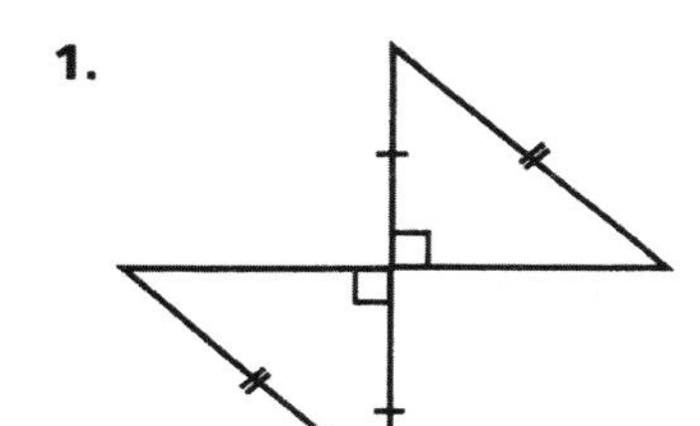

2.

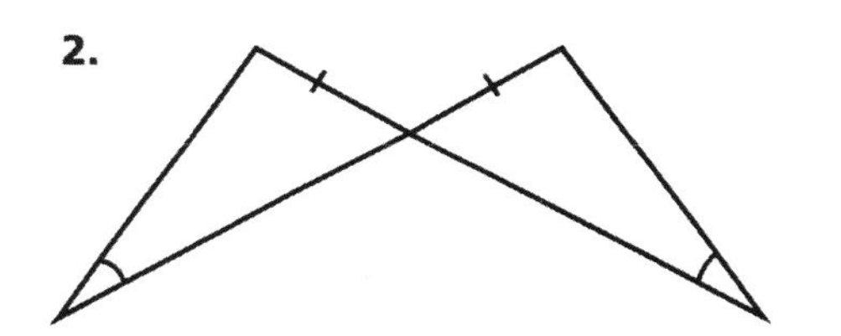

3.

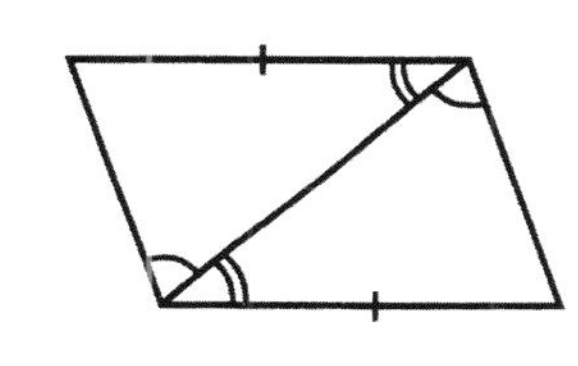

4.

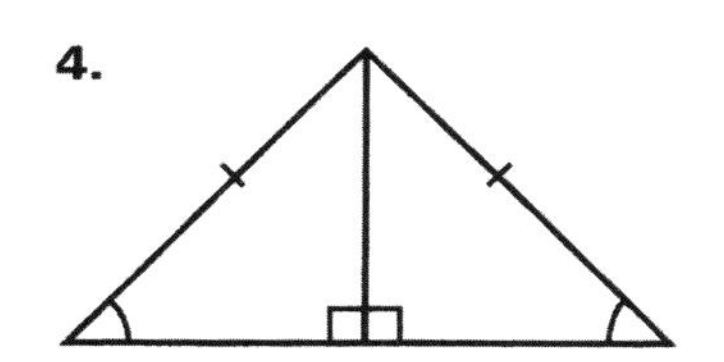

5.

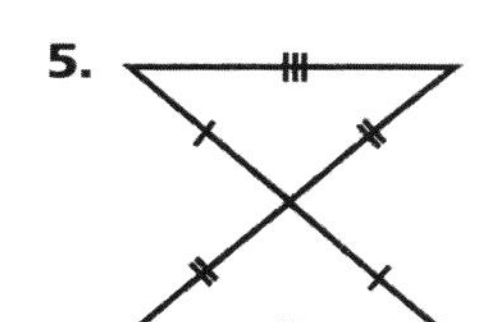

6.

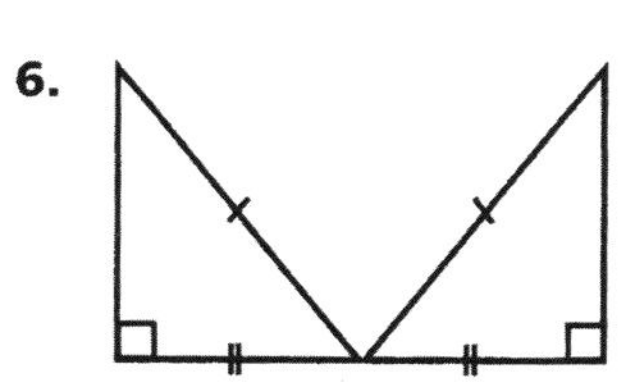

7.

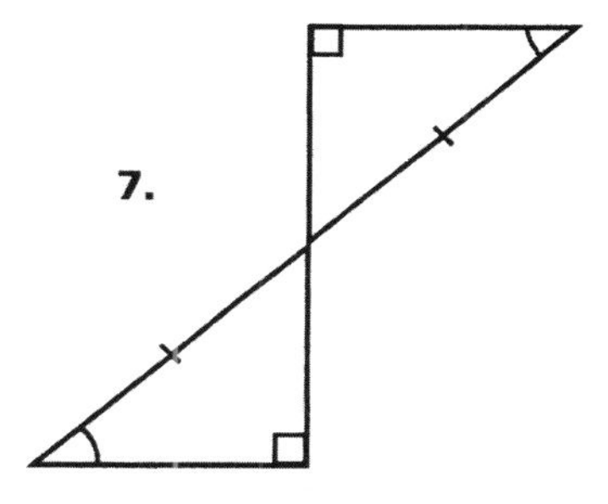

8.

9.

	SSS	SAS	ASA	AAS	HL	can't
1.	U	M	W	A	N	P
2.	B	R	I	O	L	Y
3.	A	N	E	C	Q	T
4.	M	E	P	A	N	R
5.	T	P	H	E	G	F
6.	Z	A	D	H	R	E
7.	I	M	O	V	G	U
8.	S	N	A	K	L	E
9.	W	I	T	B	E	X

How many geometry teachers does it take to change a light bulb?

_ _ _ _ . THEY _ _ _ ' _ DO IT.

THEY CAN ONLY _ _ _ _ _ _ _

_ _ CAN _ _ DONE!

© Milliken Publishing Company

Name ______________________________

Proving Congruence

TIPS! **1. By the Reflexive Property, a segment is congruent to itself. $\overline{XY} \cong \overline{XY}$**
2. This symbol →▸— indicates parallel lines.

Draw straight lines to match each statement within the proof to its reason. Each set will have an extra unused reason. The uncrossed letters will spell out a word.

1.

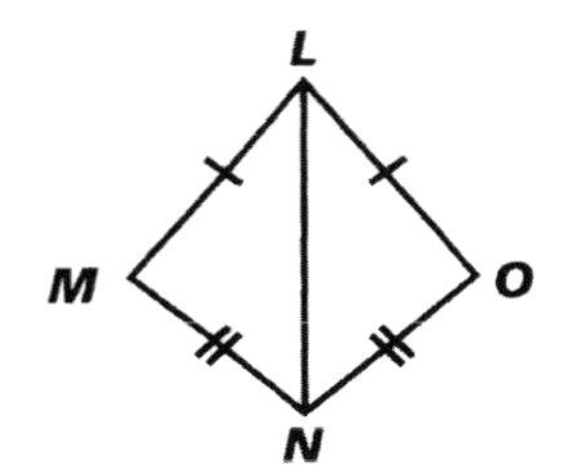

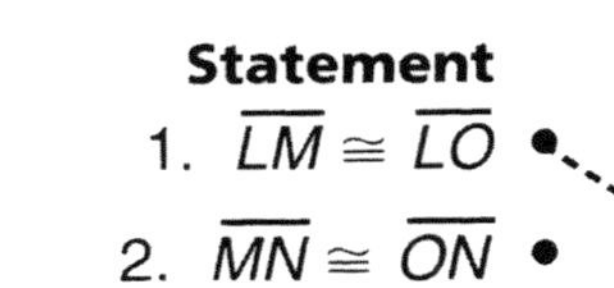

Statement		Reason
1. $\overline{LM} \cong \overline{LO}$	A	SAS Congruence
2. $\overline{MN} \cong \overline{ON}$		SSS Congruence
3. $\overline{LN} \cong \overline{LN}$	G	Given
4. $\triangle LMN \cong \triangle LON$	S	Given
		Reflexive Property

2.

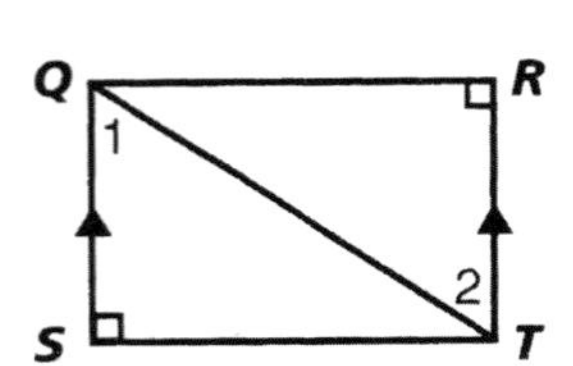

Statement		Reason
1. $\overline{QS} \parallel \overline{RT}$	W	Reflexive Property
2. $\angle R \cong \angle S$		AAS Congruence
3. $\angle 1 \cong \angle 2$	R	Alternate Interior Angles
4. $\overline{QT} \cong \overline{QT}$		SAS Congruence
5. $\triangle QST \cong \triangle TRQ$		Right Angle Congruence
	E	Given

3.

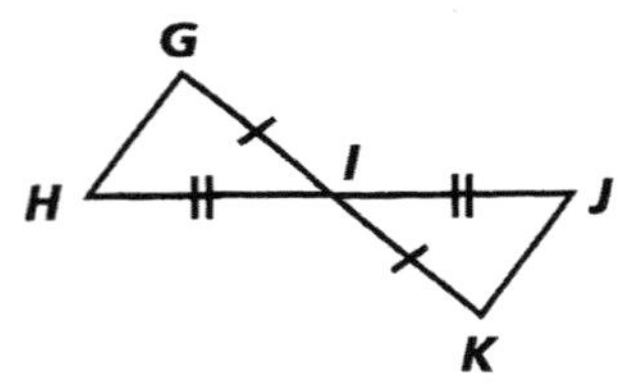

Statement		Reason
1. $\overline{GI} \cong \overline{KI}$	S	Vertical Angles
2. $\overline{HI} \cong \overline{JI}$	U	SAS Congruence
3. $\angle GIH \cong \angle KIJ$		Given
4. $\triangle GIH \cong \triangle KIJ$		Given
	O	SSS Congruence

4.

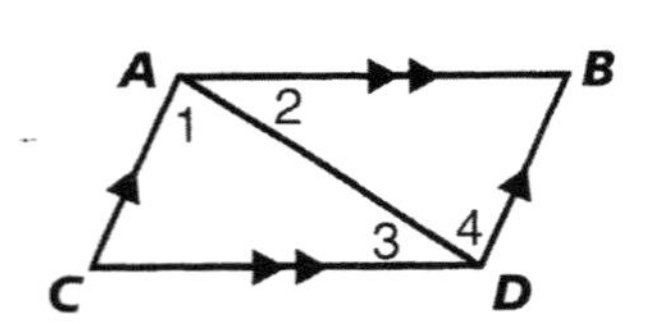

Statement		Reason
1. $\overline{AC} \parallel \overline{BD}$, $\overline{AB} \parallel \overline{CD}$	P	Alternate Interior Angles
2. $\angle 1 \cong \angle 4$, $\angle 2 \cong \angle 3$	M	AAS Congruence
3. $\overline{AD} \cong \overline{AD}$	E	Reflexive Property
4. $\triangle ADC \cong \triangle DAB$		Given
	A	ASA Congruence

5.

Statement		Reason
1. $\angle XWY$ and $\angle XWZ$ are right angles	T!	Definition of Right Triangles
2. $\triangle XWY$ and $\triangle XWZ$ are right triangles		Given
3. $\overline{XY} \cong \overline{XZ}$	E!	SAS Congruence
4. $\overline{XW} \cong \overline{XW}$		HL Congruence
5. $\triangle XWY \cong \triangle XWZ$	R!	Reflexive Property
		Given

Name ____________________

The Pythagorean Theorem

Remember

In a right triangle, the sum of the squares of the legs is equal to the square of the hypotenuse:

$c^2 = a^2 + b^2$

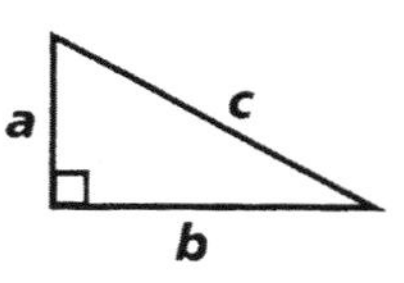

Example:

Find the length of the missing side.

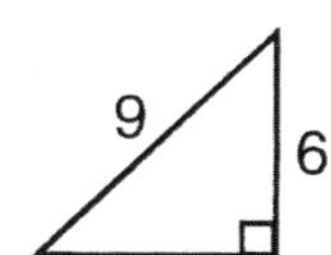

$$9^2 = 6^2 + b^2$$
$$81 = 36 + b^2$$
$$45 = b^2$$
$$\sqrt{45} = b$$
$$\sqrt{9} \cdot \sqrt{5} = b$$
$$3\sqrt{5} = b$$

Solve for the missing side. Use the answer code to find the special name for three integers whose lengths form a right triangle.

TIP! A 3-4-5 triangle has a leg-to-leg-to-hypotenuse ratio of 3:4:5. If you can spot multiples of these numbers, you can solve those problems easily.

A.

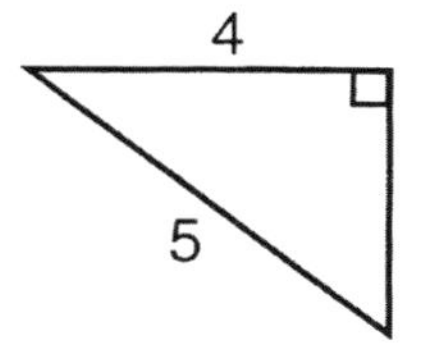

E.

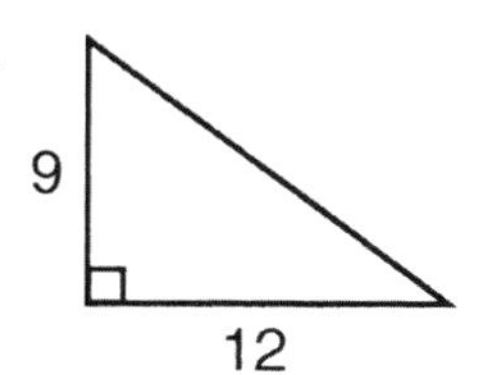

G.

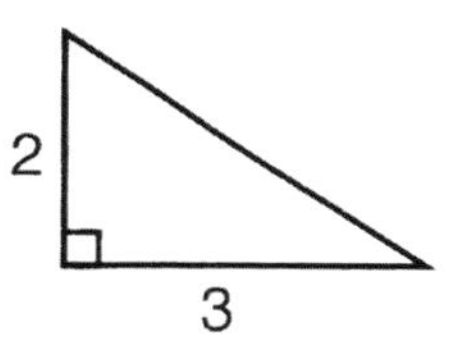

H.

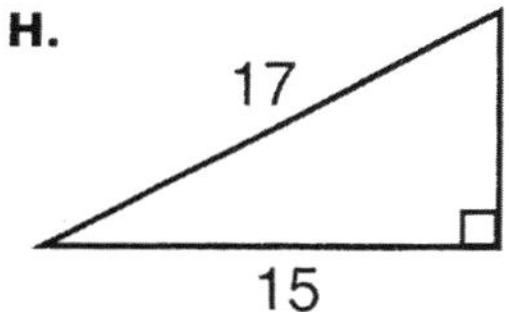

I.

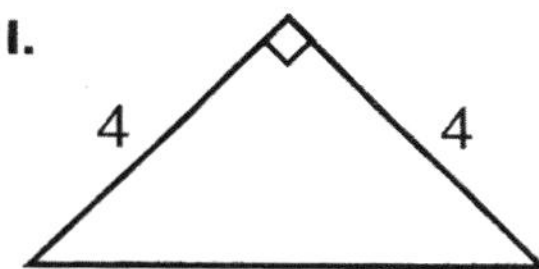

L.

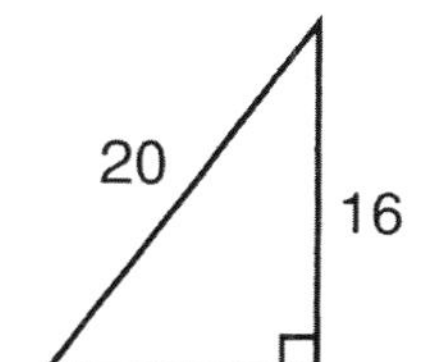

N.

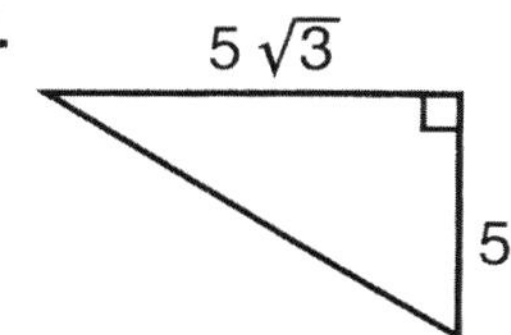

O.

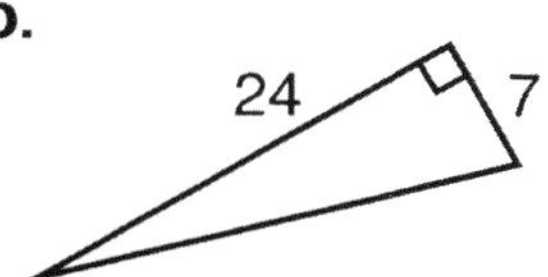

P.

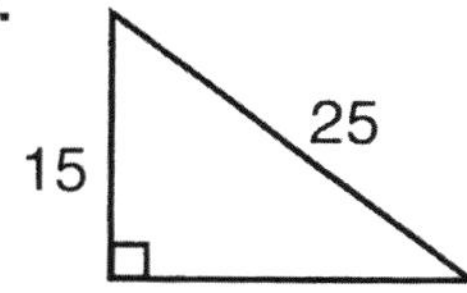

R.

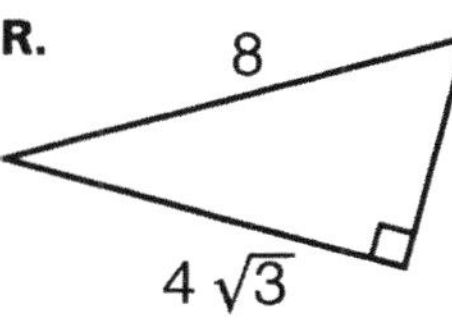

S.

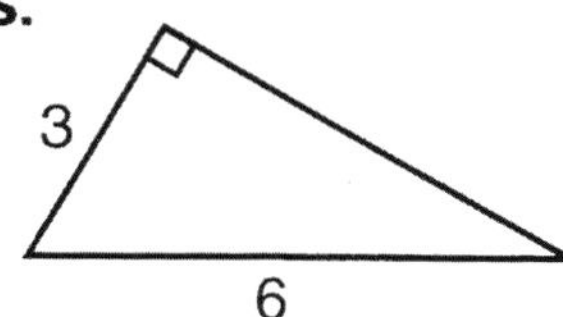

T.

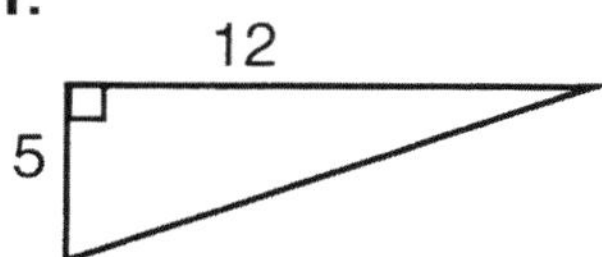

Y.

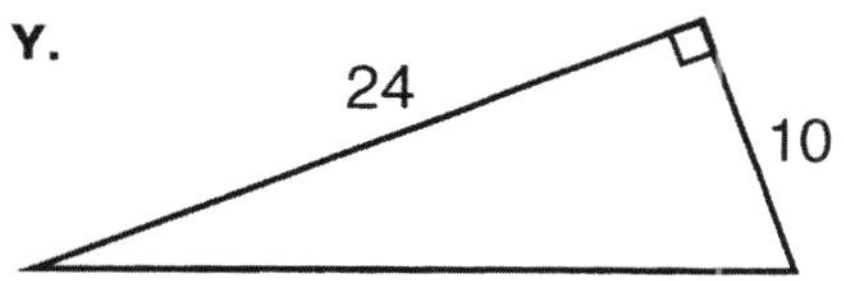

__ __ __ __ __ __ __ __ __ __ __

20 26 13 8 3 $\sqrt{13}$ 25 4 15 3 10

__ __ __ __ __ __ __

13 4 $4\sqrt{2}$ 20 12 15 $3\sqrt{3}$

© Milliken Publishing Company

Name ______________________________

The Converse of the Pythagorean Theorem

Remember

The Pythagorean Theorem can be used to determine whether a triangle is right, acute, or obtuse. Think of the long side as c and the two shorter sides as a and b.

If $c^2 = a^2 + b^2$, then it is a right triangle.

$25 = 9 + 16$

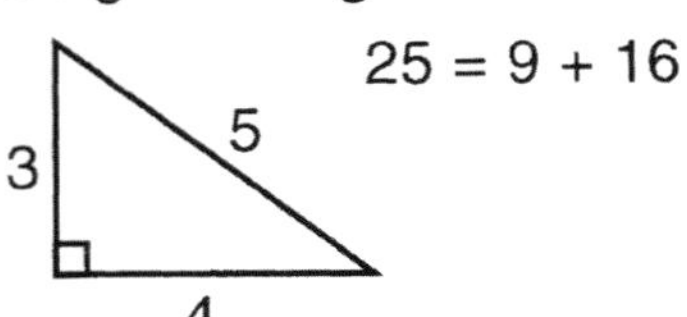

If $c^2 < a^2 + b^2$, then it is an acute triangle.

$36 < 16 + 25$

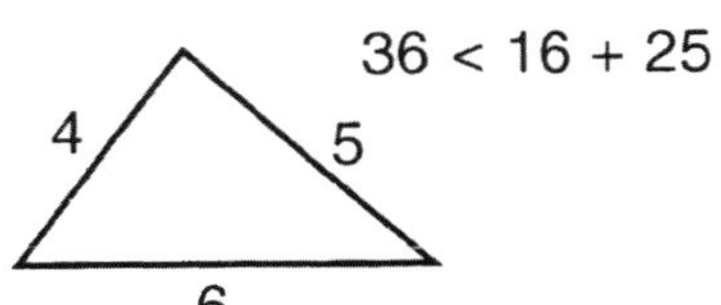

If $c^2 > a^2 + b^2$, then it is an obtuse triangle.

$16 > 9 + 4$

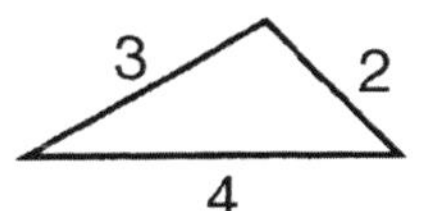

Determine whether the following lengths create a right, acute, or obtuse triangle or no triangle. Shade in the matching column letters and copy them onto the blanks to reveal the name of the U.S. president who discovered a proof of the Pythagorean Theorem.

	lengths	right	acute	obtuse	no triangle
1.	6, 8, 10	J	A	T	W
2.	1, 2, 3	E	H	B	A
3.	5, 5, 5	R	M	O	S
4.	7, 8, 12	A	M	E	H
5.	7, 8, 9	R	S	H	A
6.	5, 9, 11	O	N	A	S
7.	5, 12, 13	G	Z	A	M
8.	11, 11, 15	P	A	P	T
9.	16, 30, 34	R	E	L	O
10.	20, 40, 50	A	G	F	N
11.	9, 12, 15	I	A	S	H
12.	5, 7, 13	S	N	G	E
13.	8, 14, 17	O	N	L	Y
14.	13, 18, 22	T	D	N	H

___ ___ ___ ___ ___   ___.   ___ ___ ___ ___ ___ ___ ___ ___

© Milliken Publishing Company

Name ______________________________

The Distance Formula

Remember

To find the distance between two points, use the Distance Formula. It is based on the Pythagorean Theorem.

$$D = \sqrt{(x_2 - x_1)^2 + (y_2 - y_1)^2}$$

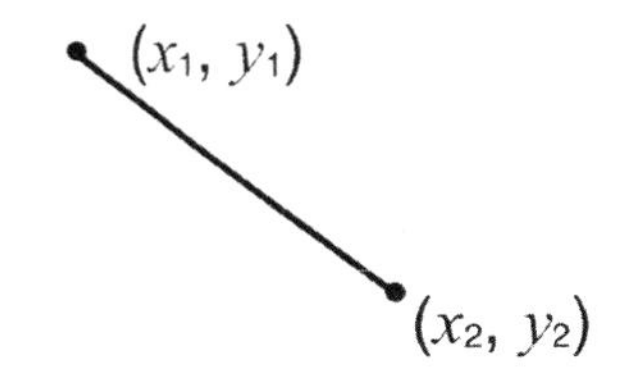

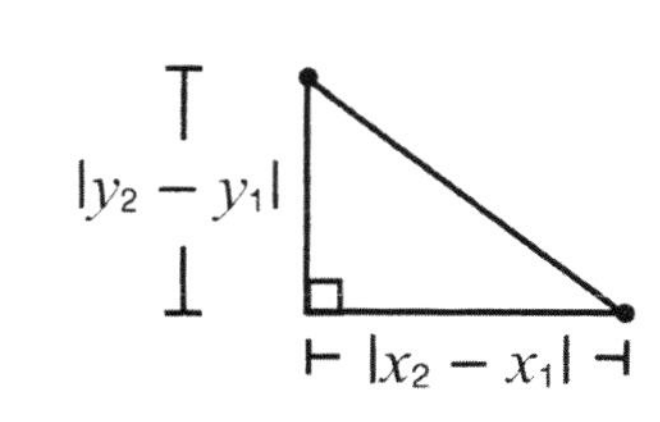

Example: Find the distance between (−11, 7) and (−9, 3). Let (−11, 7) be (x_1, y_1) and (−9, 3) be (x_2, y_2).

$$\begin{aligned} D &= \sqrt{(-9 - -11)^2 + (3 - 7)^2} \\ &= \sqrt{(2)^2 + (-4)^2} \\ &= \sqrt{4 + 16} \\ &= \sqrt{20} \\ &= \sqrt{4 \cdot 5} \\ &= 2\sqrt{5} \text{ units} \end{aligned}$$

Label each pair of points on the graph and find the distance between them. Use your answers and the decoder to find the distance from the center of the pitcher's mound to home plate in baseball.

1. ***A*** (−11, 7) and ***B*** (−9, 3) $2\sqrt{5}$
2. ***C*** (−5, −3) and ***D*** (−5, 8) ________
3. ***E*** (−7, 5) and ***F*** (1, −1) ________
4. ***G*** (8, 3) and ***H*** (3, −2) ________
5. ***I*** (−3, 10) and ***J*** (1, 7) ________
6. ***K*** (−10, −9) and ***L*** (−2, 6) ________
7. ***M*** (7, 9) and ***N*** (5, 8) ________
8. ***O*** (2, −12) and ***P*** (7, 0) ________
9. ***Q*** (−4, 5) and ***R*** (8, −4) ________
10. ***S*** (−1, −7) and ***T*** (−4, −8) ________

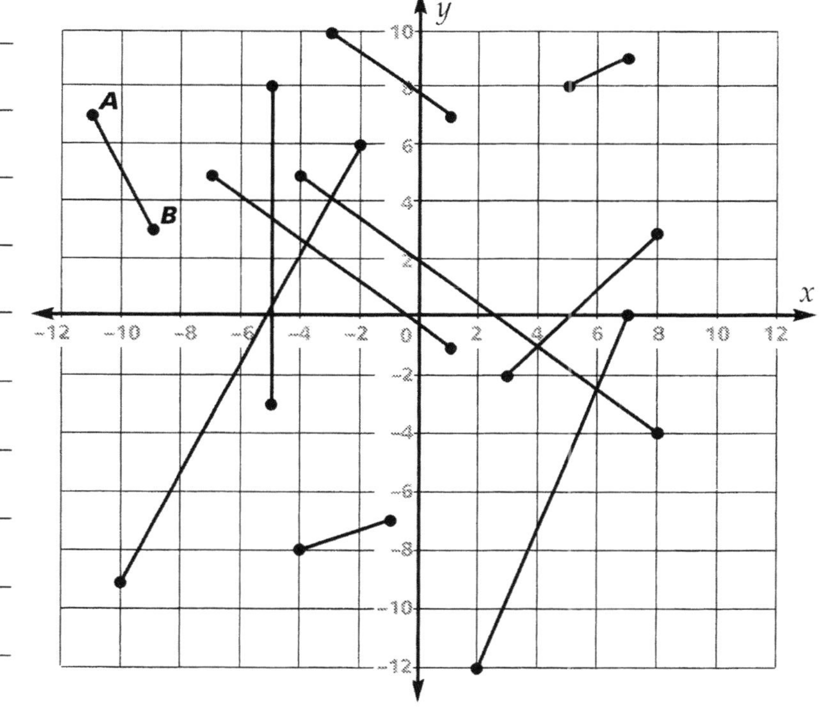

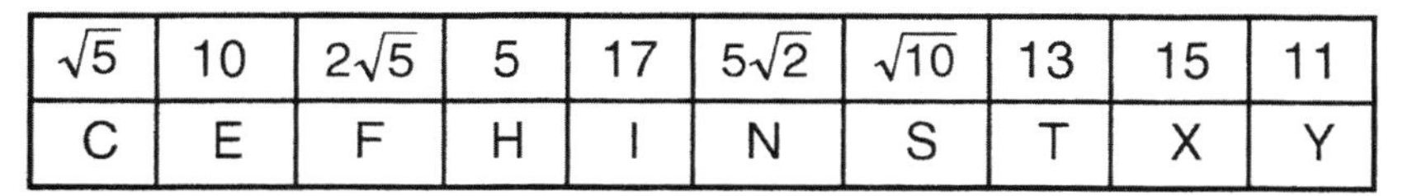

$\sqrt{5}$	10	$2\sqrt{5}$	5	17	$5\sqrt{2}$	$\sqrt{10}$	13	15	11
C	E	F	H	I	N	S	T	X	Y

__ __ __ __ __   F __ __ __ ,
10 6 9 8 2   1 3 3 8

__ __ __   __ __ __ __ __ __
10 6 9   6 4 7 5 3 10

© Milliken Publishing Company

Name ___________________________

Special Right Triangles

Remember

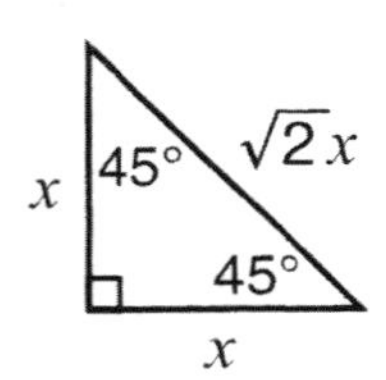

1. In a **45°–45°–90°** right triangle, the hypotenuse is $\sqrt{2}$ times as long as each leg.

x 60° $2x$ 30° $\sqrt{3}x$

2. In a **30°–60°–90°** right triangle, the hypotenuse is twice as long as the short leg. The long leg is $\sqrt{3}$ times as long as the short leg.

Example: Find the missing lengths.

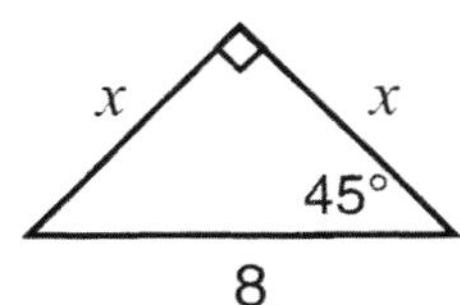

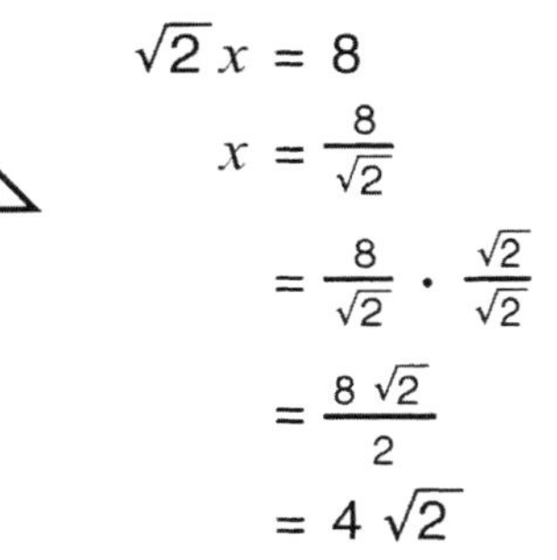

$$\sqrt{2}x = 8$$
$$x = \frac{8}{\sqrt{2}}$$
$$= \frac{8}{\sqrt{2}} \cdot \frac{\sqrt{2}}{\sqrt{2}}$$
$$= \frac{8\sqrt{2}}{2}$$
$$= 4\sqrt{2}$$

Use the 30°–60°–90° and the 45°–45°–90° triangle relationships to solve for the missing sides. Follow your answers in alphabetical order through the maze.

1.

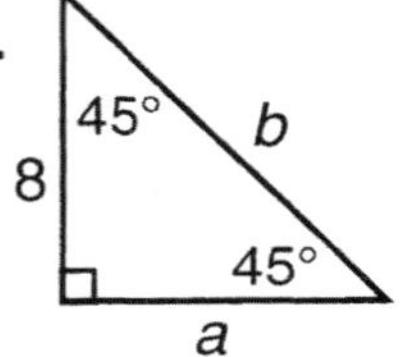

2.

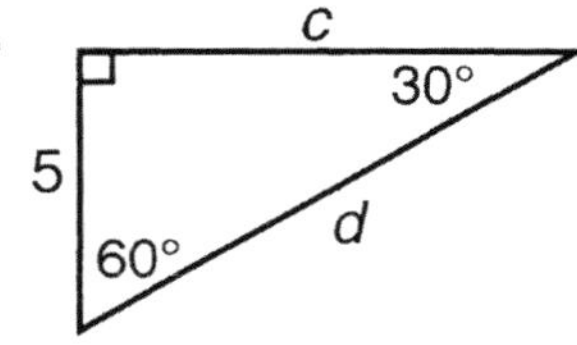

3.

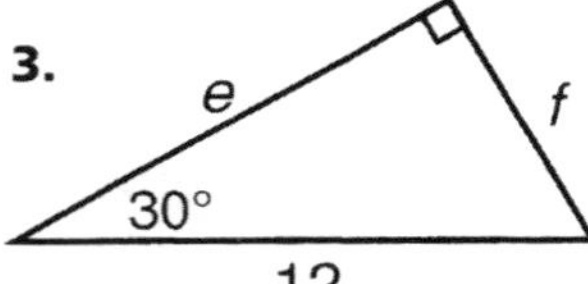

4.

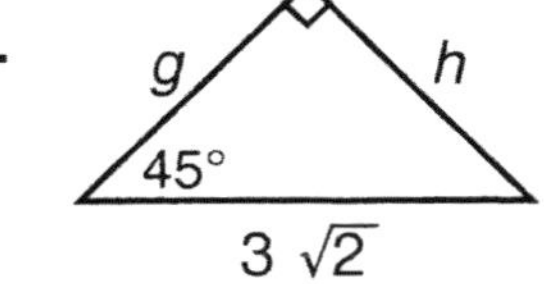

5.

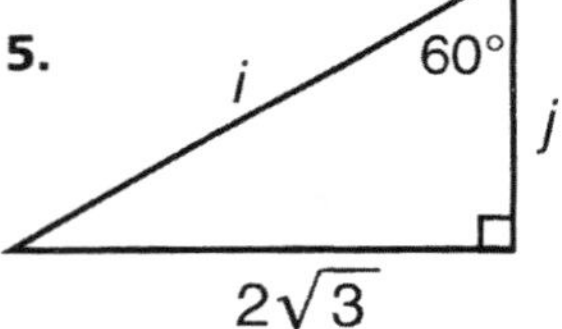

6.

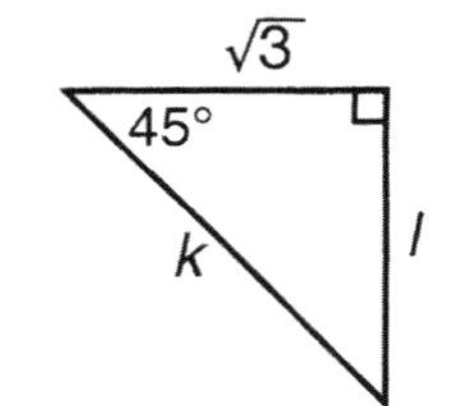

7.

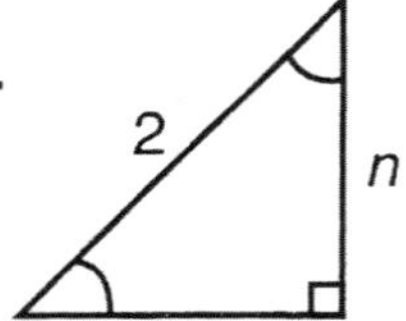

8.

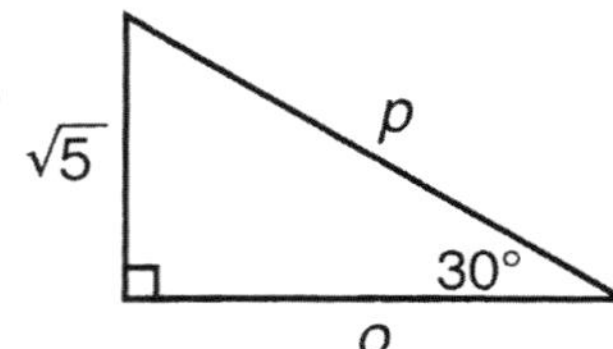

9.

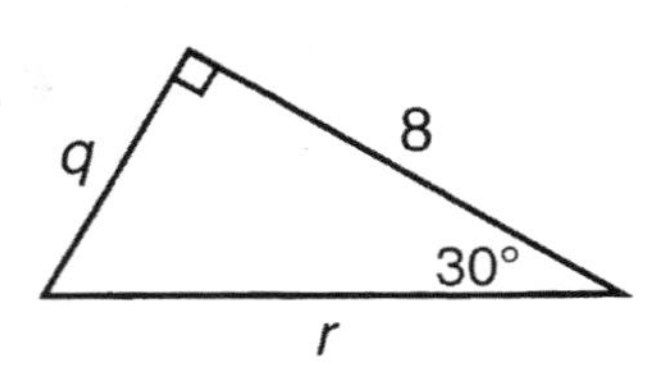

10.

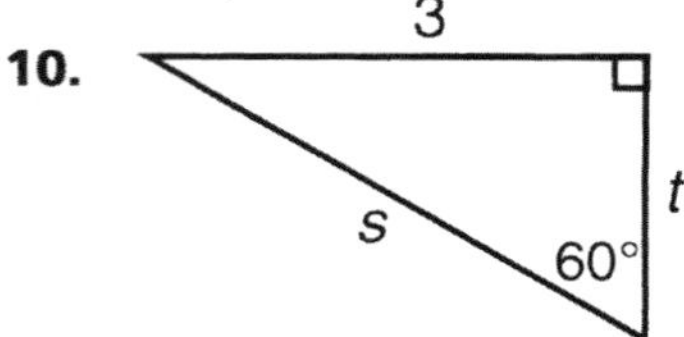

$\sqrt{6}$ 3 $5\sqrt{2}$ 6 $6\sqrt{3}$ 3 10 $5\sqrt{3}$ $6\sqrt{2}$ 1 16 $8\sqrt{2}$ $\sqrt{6}$ 4 8 2 4 $\sqrt{2}$ $8\sqrt{3}$ $\sqrt{3}$ $\sqrt{15}$ $\sqrt{2}$ 2 16 $2\sqrt{5}$ 10 4 $2\sqrt{3}$ $\frac{8\sqrt{3}}{3}$ 3 $\sqrt{3}$ $3\sqrt{3}$ $\frac{16\sqrt{3}}{3}$

© Milliken Publishing Company

Name ______________________________

Trigonometric Ratios

Remember

A trigonometric ratio is a ratio between two sides of a right triangle. *Sine, cosine,* and *tangent* are the three basic ratios. They are abbreviated as *sin, cos,* and *tan.* The made-up name "soh cah toa" can help you memorize the three basic ratios.

$$\textbf{sin} = \frac{\textbf{o}\text{pposite}}{\textbf{h}\text{ypotenuse}}$$

$$\textbf{cos} = \frac{\textbf{a}\text{djacent}}{\textbf{h}\text{ypotenuse}}$$

$$\textbf{tan} = \frac{\textbf{o}\text{pposite}}{\textbf{a}\text{djacent}}$$

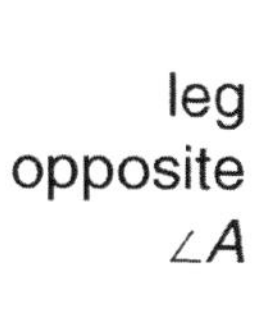

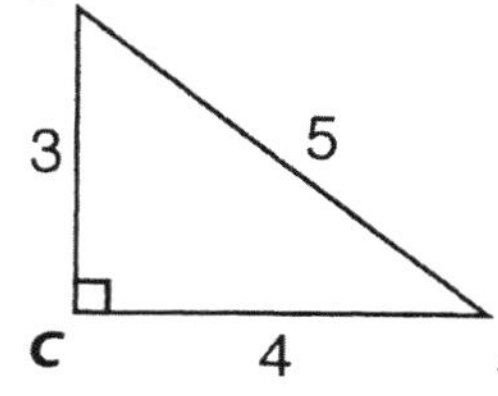

$$\sin A = \frac{\text{opp.}}{\text{hyp.}} = \frac{3}{5}$$

$$\cos A = \frac{\text{adj.}}{\text{hyp.}} = \frac{4}{5}$$

$$\tan A = \frac{\text{opp.}}{\text{adj.}} = \frac{3}{4}$$

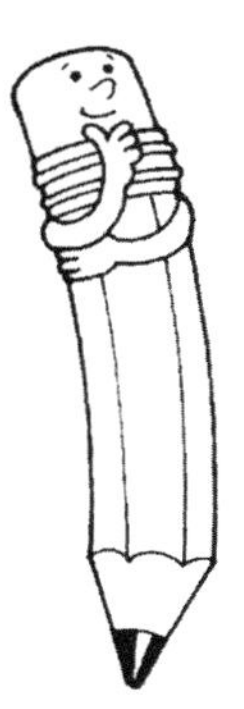

Draw straight lines to match each trigonometric ratio to its value. The uncrossed words will reveal a message.

1.

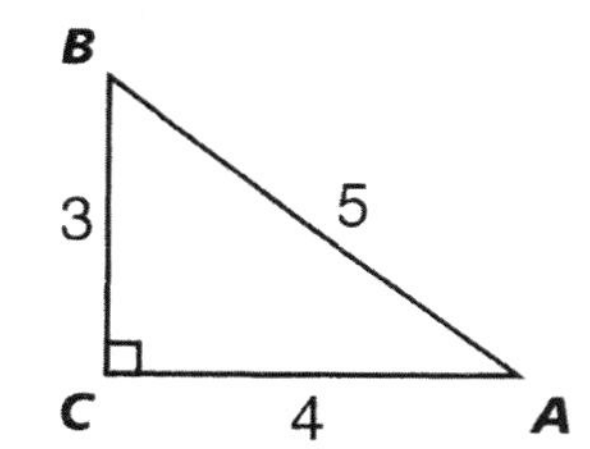

Ratio	Words	Value
sin *B* •	**You're**	• $\frac{\text{adj.}}{\text{hyp.}} = \frac{3}{5}$
cos *B* •	**Get** **You**	• $\frac{\text{opp.}}{\text{adj.}} = \frac{4}{3}$
tan *B* •		• $\frac{\text{opp.}}{\text{hyp.}} = \frac{4}{5}$

2.

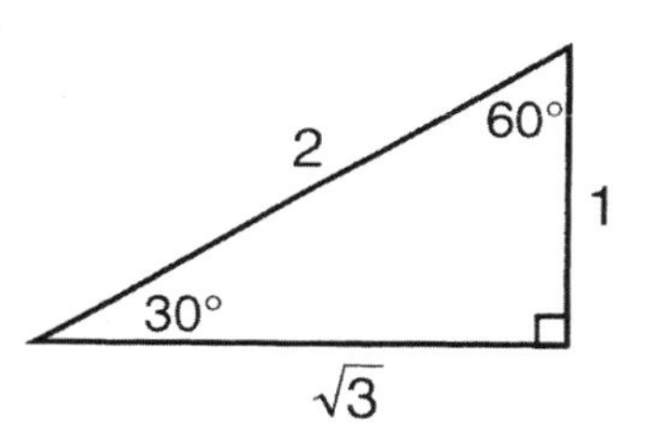

Ratio	Words	Value
sin 30° •		• $\frac{\sqrt{3}}{2}$
sin 60° •	**need** **a**	• $\frac{\sqrt{3}}{1} = \sqrt{3}$
tan 30° •	**great**	• $\frac{1}{\sqrt{3}} = \frac{\sqrt{3}}{3}$
tan 60° •		• $\frac{1}{2}$

3.

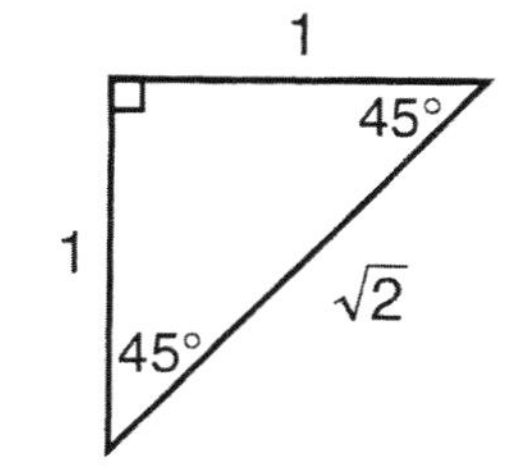

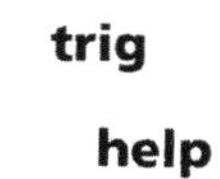

Ratio	Words	Value
cos 45° •	**trig**	• $\frac{\text{opp.}}{\text{hyp.}} = \frac{1}{\sqrt{2}} = \frac{\sqrt{2}}{2}$
tan 45° •	**help** **math**	• $\frac{\text{adj.}}{\text{hyp.}} = \frac{1}{\sqrt{2}} = \frac{\sqrt{2}}{2}$
sin 45° •		• $\frac{1}{1} = 1$

4.

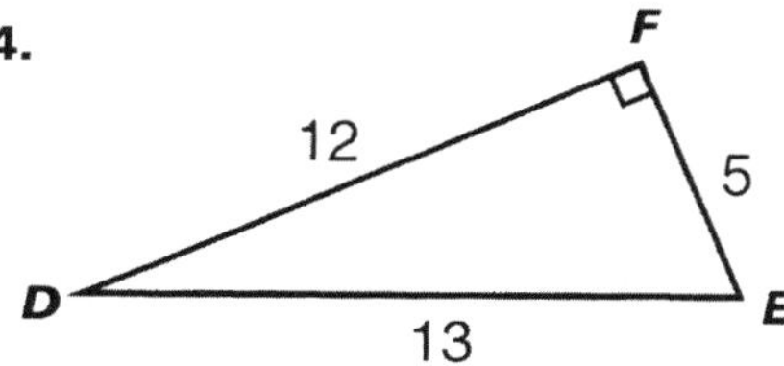

Ratio	Words	Value
tan *D* •		• $\frac{12}{13}$
tan *E* •	**student!**	• $\frac{12}{5}$
cos *D* •	**expert!**	• $\frac{5}{12}$

© Milliken Publishing Company

Name ______________________

Using Trigonometric Ratios

Example

A water slide has an angle of elevation from its base to its height of 20°. The slide has a height of 36 feet. About how long is the slide?

1. Draw and label a right triangle sketch.

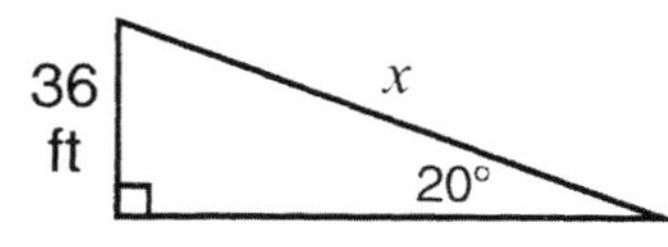

2. Decide which trigonometric ratio can help you solve the problem. $\sin 20° = \frac{\text{opp.}}{\text{hyp.}} = \frac{36}{x}$

3. Use a calculator or table to find a decimal approximation for the ratio. $\sin 20° \approx .3420$

4. Solve for *x*.

$.3420 = \frac{36}{x}$

$.3420x = 36$

$x = \frac{36}{.3420}$

$x = 105.263$

$x \approx 105$ feet

Read each problem and draw a line to its matching sketch. Write a trigonometric ratio that will help you solve the problem. Use a decimal approximation from the table and solve for x. When you finish, find and circle your answer in the box below.

Angle	Sin	Cos	Tan
15°	.2588	.9659	.2679
20°	.3420	.9397	.3640
25°	.4226	.9063	.4663
30°	.5000	.8660	.5774
50°	.7660	.6428	1.1918
65°	.9063	.4226	2.1445

1. A slide has an angle of elevation of 25°. It is 60 feet from the end of the slide to the stairway beneath the top of the slide. About how long is the slide?
2. A forester is standing 150 feet away from a tree. She measures the angle of elevation from where she is to the top of the tree as 30°. About how tall is the tree?
3. A moving sidewalk takes zoo visitors up a hill. The sidewalk rises 48 feet. Its angle of elevation is 15°. About how long is the sidewalk?
4. A dog is on the ground watching a cat on a 6-foot-high wall. The angle of elevation from the dog to the cat is 50°. About how far from the wall is the dog?
5. Micah's kite string is 72 feet long. The angle of elevation from Micah's position on the ground to the kite in the air is 65°. About how high is the kite?

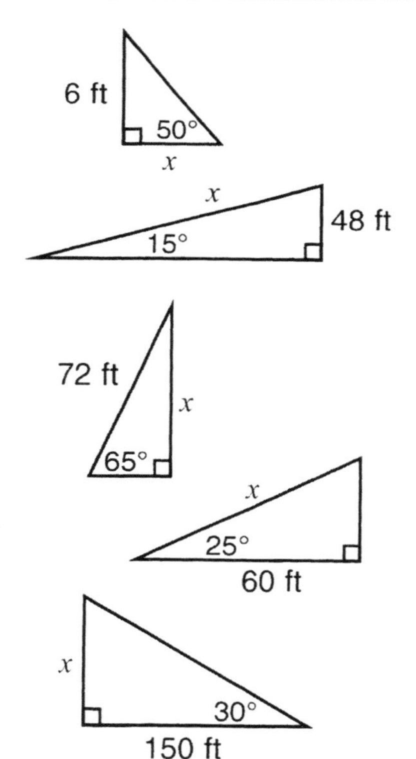

5 ft	65 ft	66 ft	87 ft	105 ft	185 ft

© Milliken Publishing Company

Name ______________________________

Angles in Polygons

Remember

A polygon is *convex* if a segment connecting any two points in the interior of the polygon is completely in the interior.

A polygon that is not convex is *concave*.

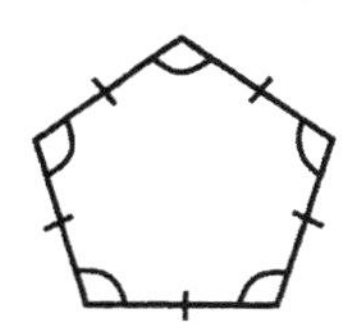

A *regular* polygon is equilateral and equiangular.

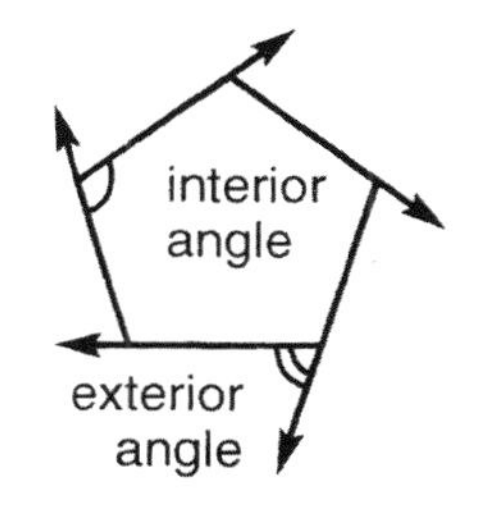

Divide each polygon below into triangles by drawing diagonals from one vertex. Then complete the chart.

Polygon	Number of sides n	Number of triangles $n - 2$	Interior angle sum of a convex polygon $(n - 2) \cdot 180°$	Measure of each interior angle in a regular polygon $\frac{(n - 2) \cdot 180°}{n}$	Exterior angle sum of a convex polygon 360°	Measure of each exterior angle in a regular polygon $\frac{360°}{n}$
triangle					360°	
quadrilateral					360°	
pentagon	5	3	$3 \cdot 180° =$ 540°	$540° \div 5 =$ 108°	360°	$360° \div 5 =$ 72°
hexagon					360°	
octagon					360°	
decagon					360°	
dodecagon					360°	

© Milliken Publishing Company

Name ______________________________

Properties of Parallelograms

Remember

1. A **parallelogram** is a quadrilateral (four-sided polygon) with these properties:
   - its opposite sides are parallel
   - its opposite sides are congruent
   - its opposite angles are congruent
   - its diagonals bisect each other

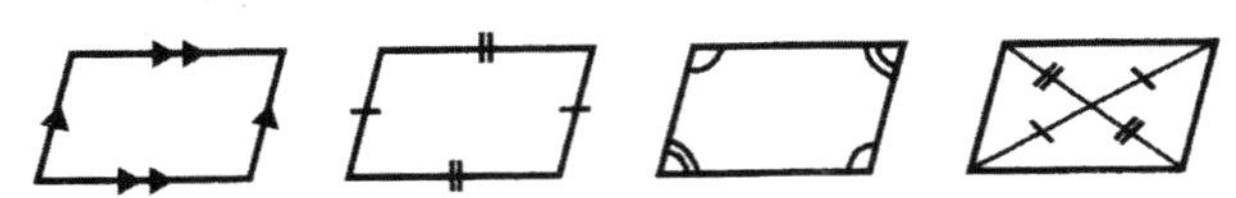

2. A **rhombus** is a parallelogram with four congruent sides. Its diagonals are perpendicular to each other and bisect opposite angles.

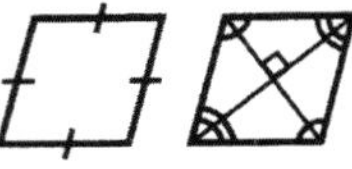

3. A **rectangle** is a parallelogram with four right angles. Its diagonals are congruent.

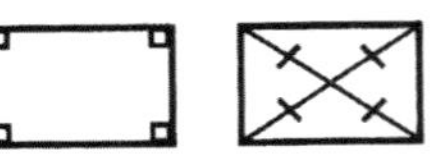

4. A **square** is a parallelogram with four congruent sides and four right angles. A square is both a rhombus and a rectangle.

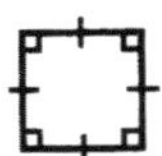

Use the properties to solve for the missing measures. Shade your answers below to reveal the answer to this riddle: *What keeps a square from moving?*

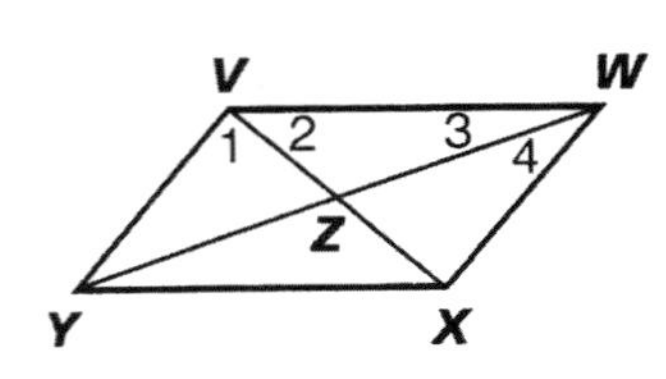

1. Given: $VWXY$ is a parallelogram. $VW = 14$ $WX = 9$ $VZ = 5.5$ $m\angle VYX = 52°$ $m\angle 2 = 40°$ $m\angle 3 = 20°$

   a. $XY =$ ______ c. $m\angle VWX =$ ______ e. $m\angle 4 =$ ______

   b. $VX =$ ______ d. $m\angle YVW =$ ______ f. $m\angle 1 =$ ______

2. Given: $PQRS$ is a rhombus. $PQ = 4$ $m\angle PQR = 60°$

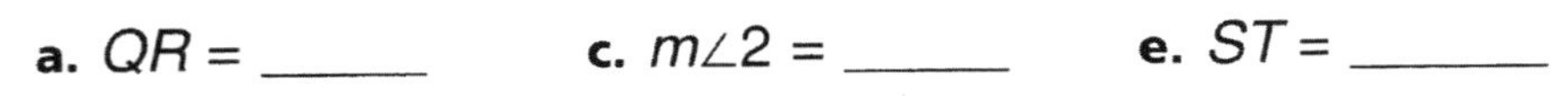

   a. $QR =$ ______ c. $m\angle 2 =$ ______ e. $ST =$ ______

   b. $m\angle 3 =$ ______ d. $PT =$ ______ f. $m\angle SPQ =$ ______

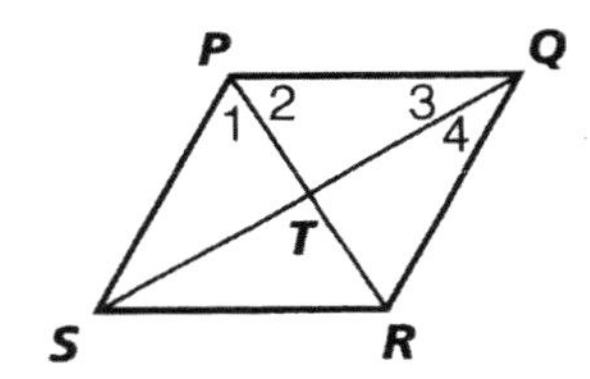

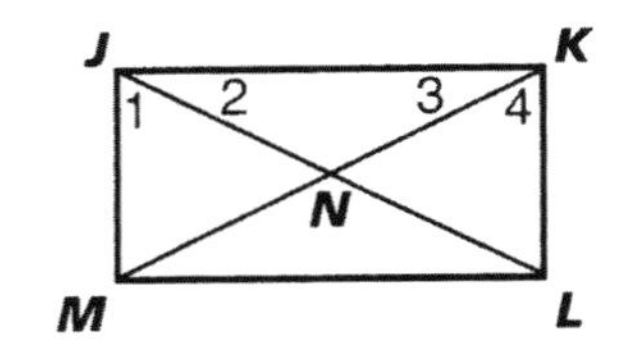

3. Given: $JKLM$ is a rectangle. $JK = 16$ $KL = 12$ $m\angle 1 = 53°$

   a. $m\angle JKL =$ ______ c. $m\angle 2 =$ ______ e. $m\angle JNK =$ ______

   b. $JL =$ ______ d. $m\angle 4 =$ ______ f. $MN =$ ______

4. Given: $ABCD$ is a square. $AB = 8$

   a. $BC =$ ______ c. $m\angle 2 =$ ______ e. $EC =$ ______

   b. $m\angle ABC =$ ______ d. $AC =$ ______ f. $m\angle BDC =$ ______

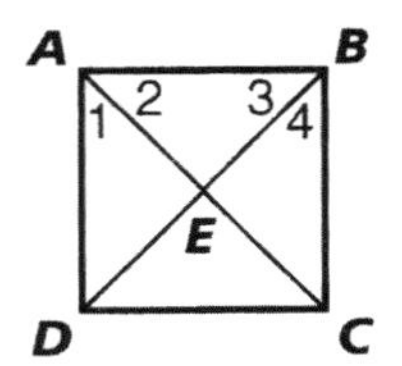

52° 32° 14 11 88° 128° 70° 9° $3\sqrt{5}$ 3 4 7 60° 0 $2\sqrt{3}$ 2 10° $12\sqrt{3}$ $\sqrt{5}$ 144° 30° 90° 37° 120° 1 25° 5° 50° 53° 20 10 106° 9 19 180° 45° 45° 90° $7\sqrt{3}$ 136° $\sqrt{5}$ $8\sqrt{2}$ 8 $4\sqrt{2}$ 135° 5 $5\sqrt{2}$

© Milliken Publishing Company

Name ________________________________

Trapezoids

Remember

1. A **trapezoid** is a quadrilateral with exactly one pair of parallel sides. The parallel sides are the bases and the nonparallel sides are the legs.
2. An **isosceles trapezoid** has congruent legs, base angles, and diagonals.
3. A **right trapezoid** has two right angles.
4. Any trapezoid can be divided into a rectangle and triangle(s) by drawing altitudes between the bases.

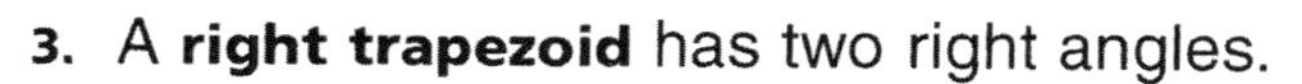

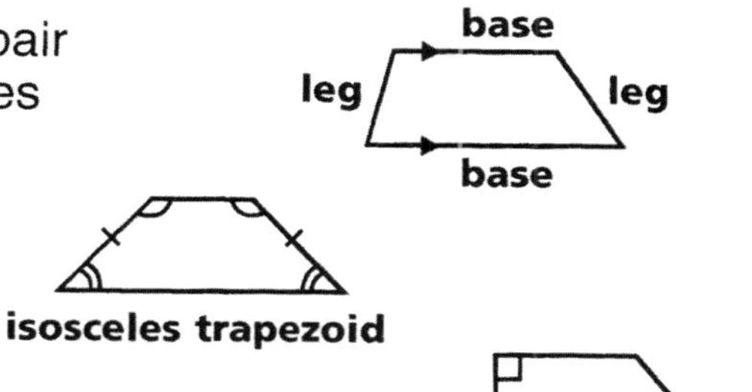

Use the properties of trapezoids, rectangles and right triangles to find the missing measures. Shade the answers below to find which U.S. state resembles a right trapezoid.

1.

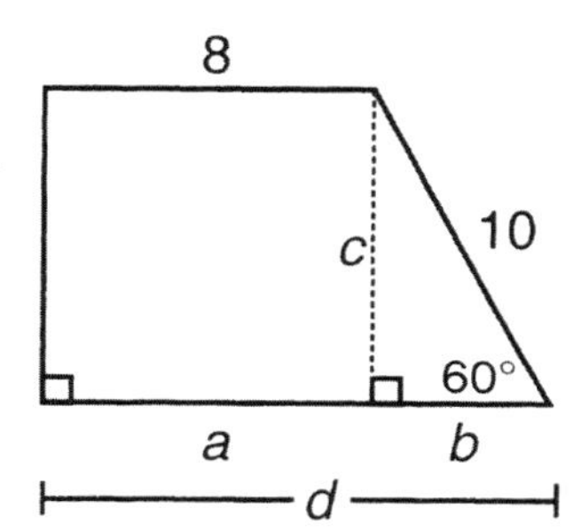

2.

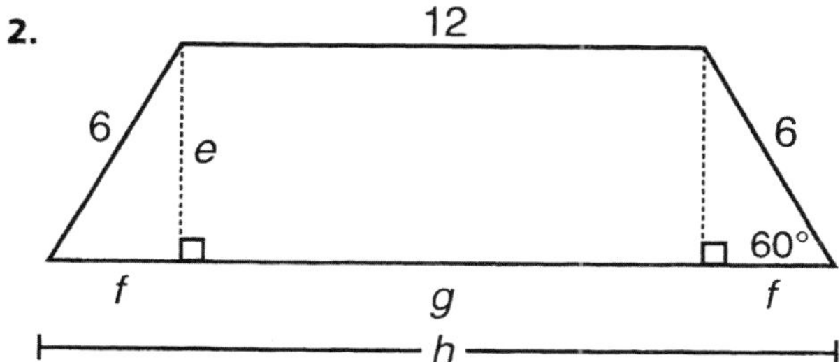

3.

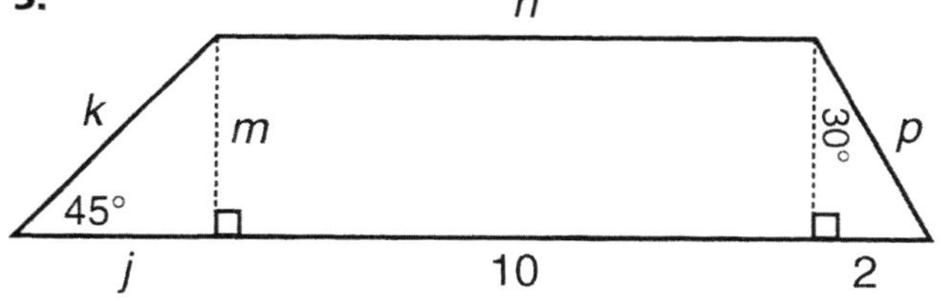

4.

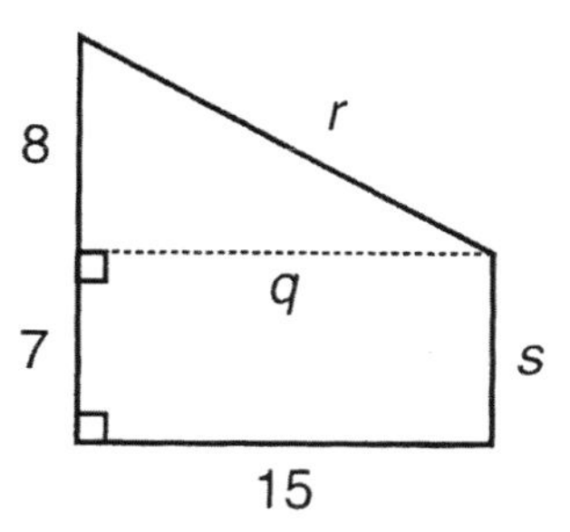

5.

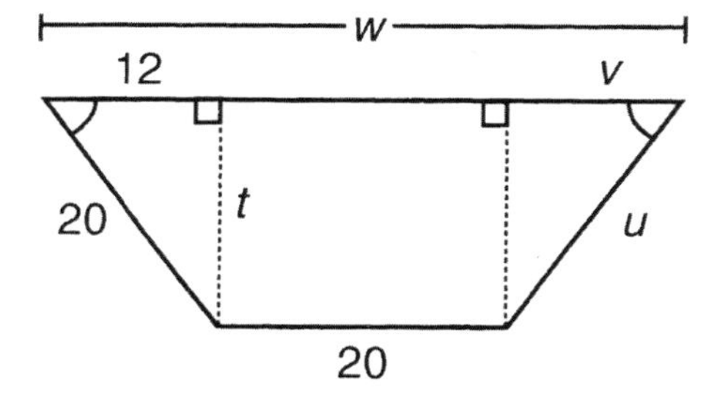

6.

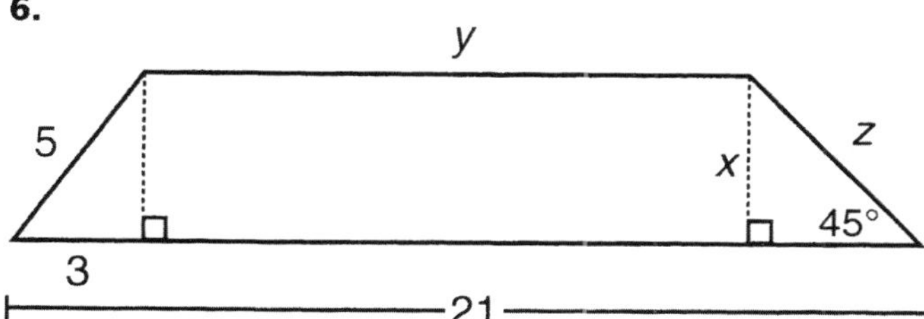

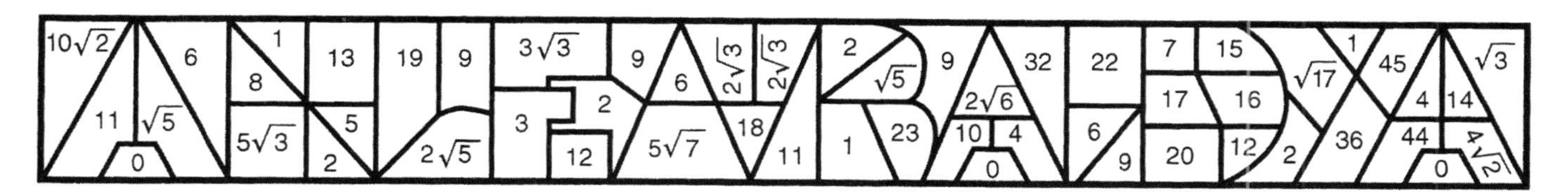

© Milliken Publishing Company

Name ______________________________

Similar Figures

Remember

1. An equation that sets two ratios equal to one another is a *proportion*.
2. Two figures are **similar** if their corresponding angles are congruent and the lengths of their corresponding sides are proportional.

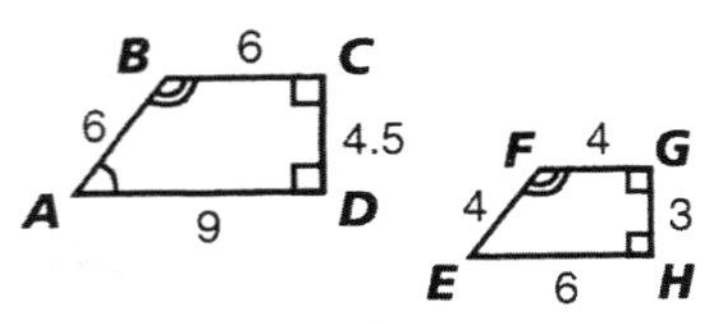

$\square ABCD \sim \square EFGH$

↑

is similar to

Example: These triangles are similar. Find the missing length.

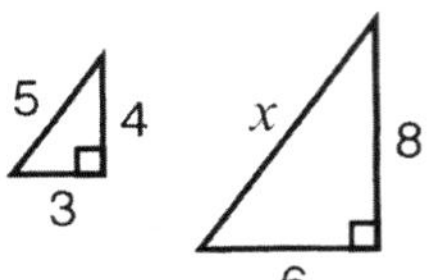

$\frac{3}{6} = \frac{5}{x}$ **Set up a proportion with corresponding sides.**

$\frac{3}{6} \times \frac{5}{x}$ **Cross multiply.**

$3x = 30$

$x = 10$ **Divide to solve for the variable.**

Use proportions to solve for the missing sides in the similar figures below. Then use the answer code to reveal the mathematician who developed the symbol for similarity ~.

1.

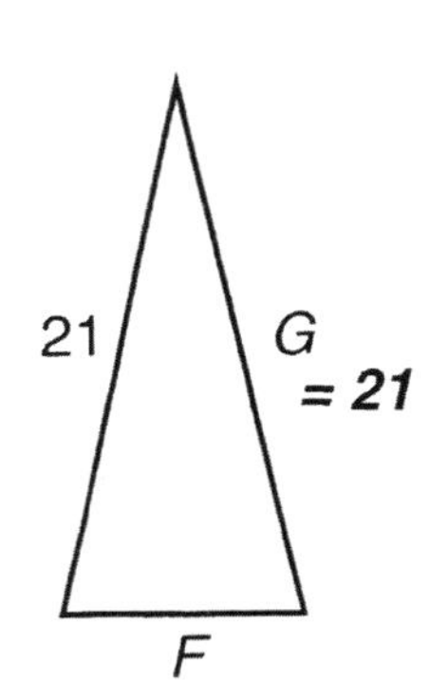

2.

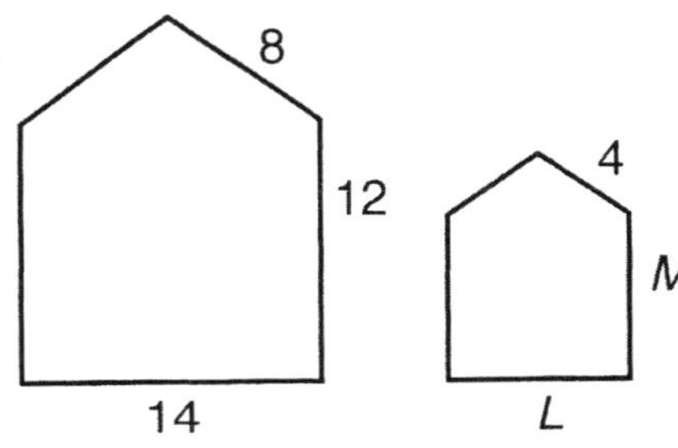

3.

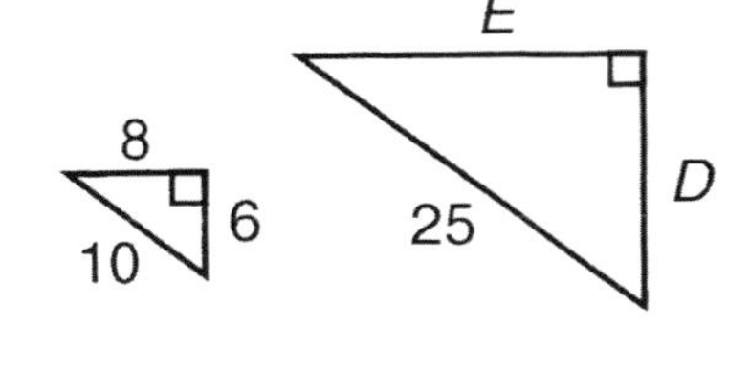

4.

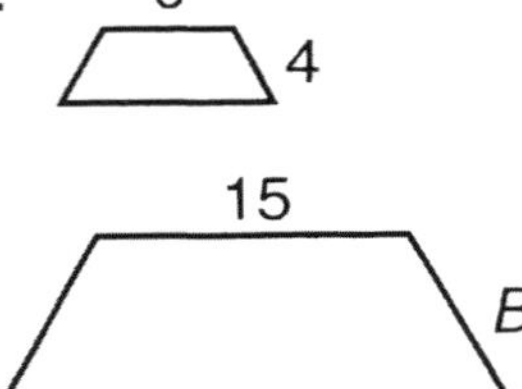

5.

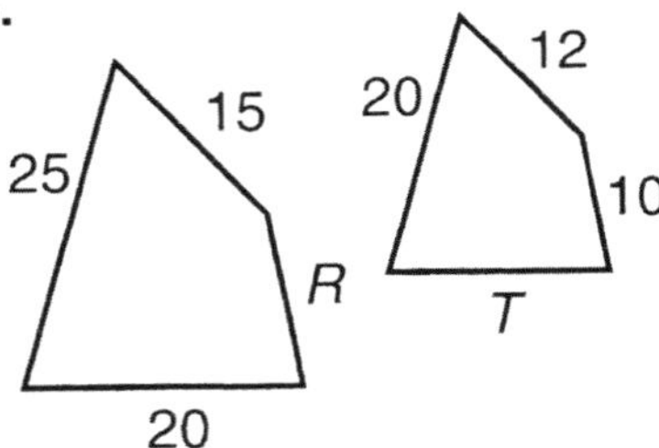

6.

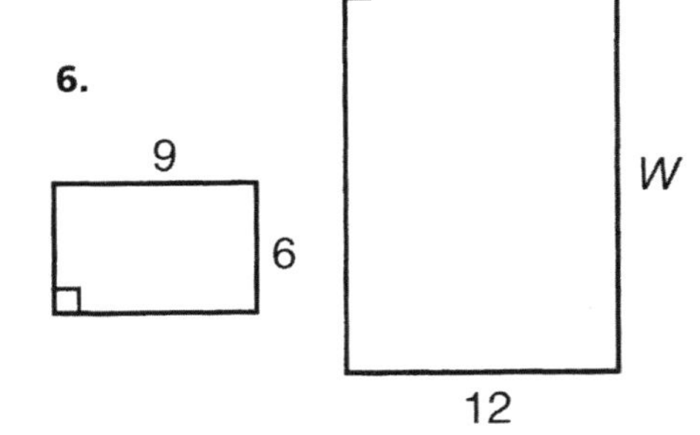

7.

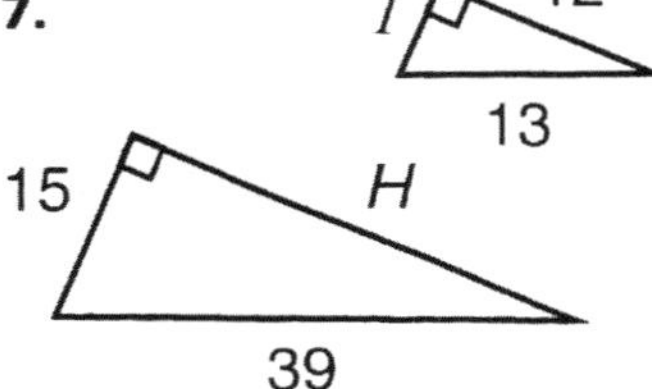

8.

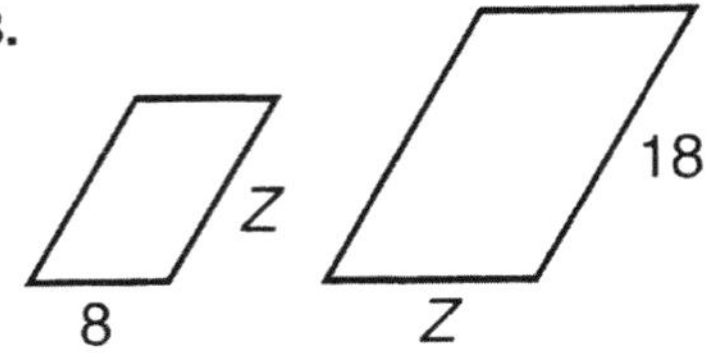

9.

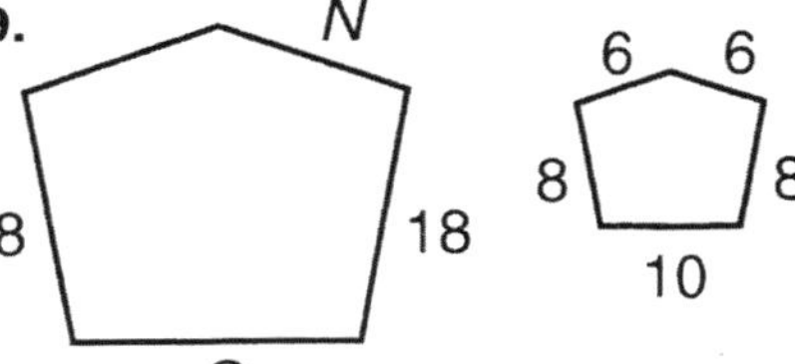

G __ __ __ __ __ __ __ __

21 22.5 16 16 9 12.5 5 20 15

__ __ __ __ __ __ __

18 5 7 36 20 7 6

__ __ __ __ __ __ __

7 20 5 10 13.5 5 12

© Milliken Publishing Company

Name ______________________________

Perimeter

Remember

Perimeter is the distance around a figure. It is measured in linear units, such as inches or centimeters. To find the perimeter of a figure, add the lengths of its sides.

Example: Find the perimeter.

4 cm

6 cm

7 cm

4 + 4 + 6 + 6 + 7 = 27 cm

Find the perimeter of each figure. Place your answers in the cross-number puzzle.

Across

1. A regular decagon with sides of 8 m.
2. A regular pentagon with sides of 9 cm.
6. A regular dodecagon with sides of 5 m.
7. A rectangle with a diagonal measure of 13 m and a side measure of 5 m.
8. A rhombus with a side of 40 ft.
9. A square whose diagonal measures $2\sqrt{2}$ m.
10. A right triangle with legs of 3 in. and 4 in.
12. A regular hexagon with sides of 8 mm.

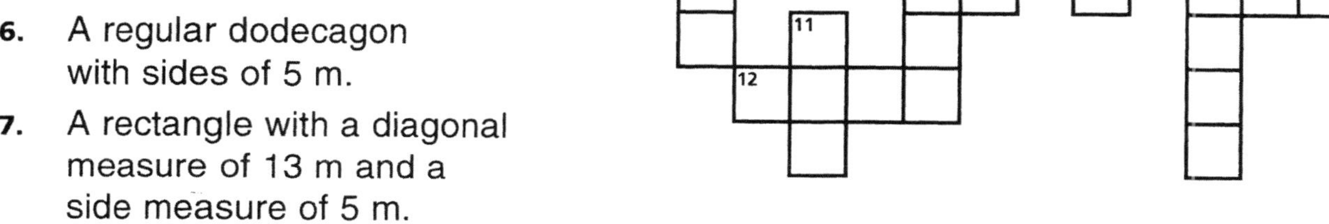

Down

1.
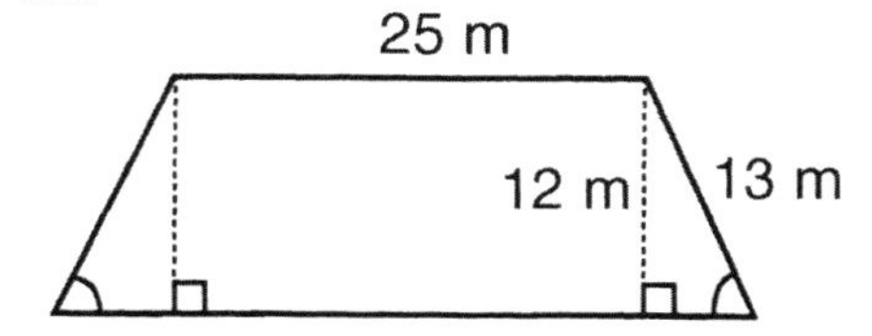

2.
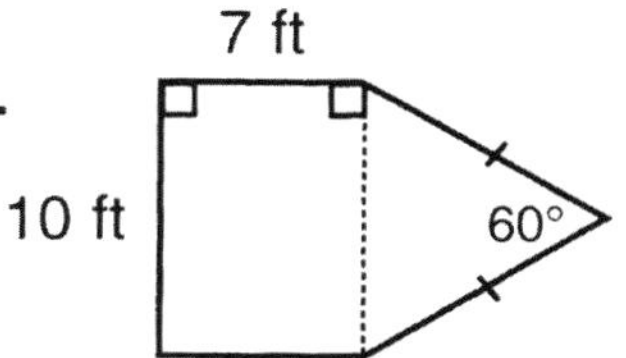

3.
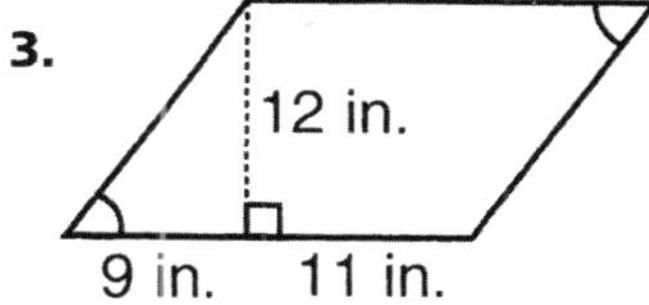

4.
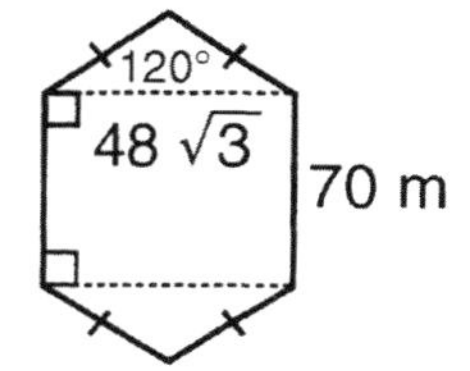

5.
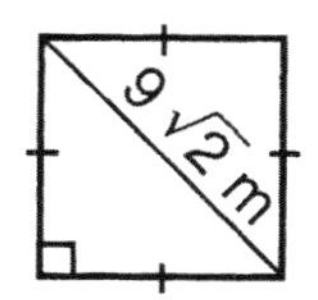

8. An equilateral triangle with sides of 6 cm.

10. A regular octagon with with sides of 2 cm.

11. A right trapezoid whose parallel sides are 15 m apart and measure 9 m and 17 m.

© Milliken Publishing Company

Name ______________________________

Area of Squares, Rectangles, and Triangles

Remember

The **area** of a figure is the number of square units needed to fill its interior.

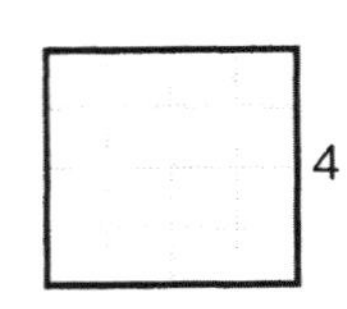

Square: $A = s^2$

Area = side x side

$A = 4^2 = 16$ units^2

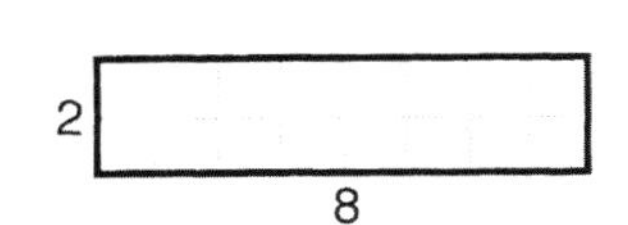

Rectangle: $A = lw$

Area = length x width

$A = 8 \cdot 2 = 16$ units^2

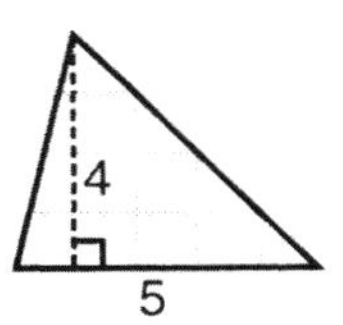

Triangle: $A = \frac{1}{2}bh$

Area = $\frac{1}{2}$ x base x height

$A = \frac{1}{2} \cdot 5 \cdot 4 = 10$ units^2

Height must be perpendicular to the base.

Draw straight lines to match each figure to its area in the center column. The uncrossed letters will spell out a message.

14

7

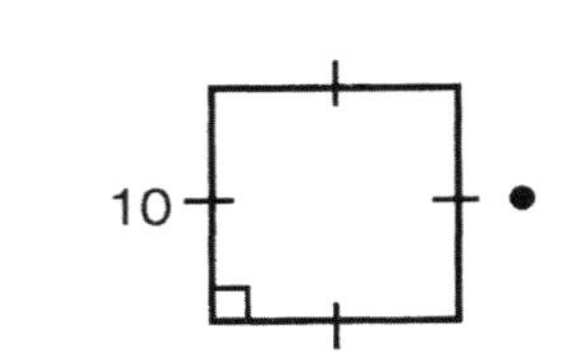

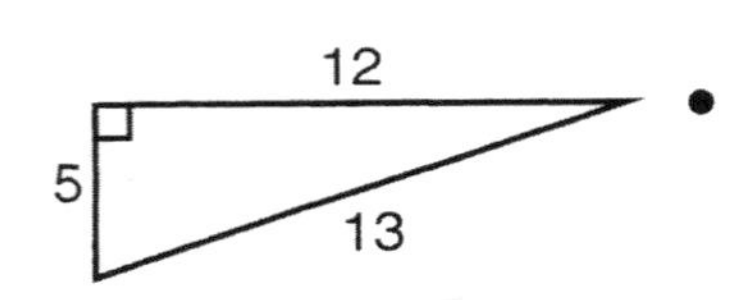

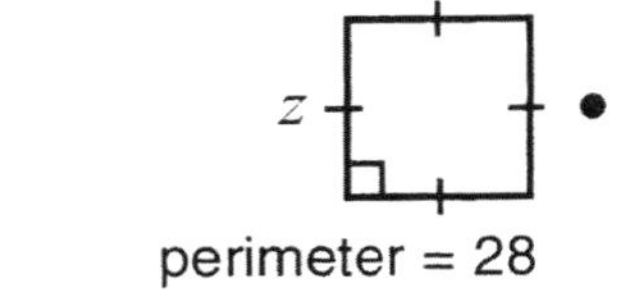

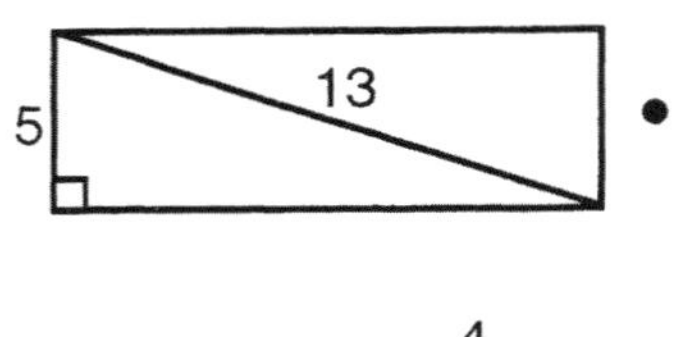

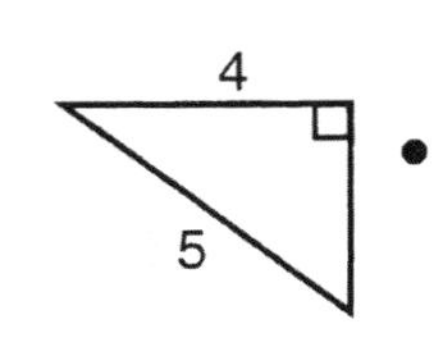

Area

- 100 units^2
- 50 units^2
- 60 units^2
- 98 units^2
- $18\sqrt{3}$ units^2
- 30 units^2
- $64\sqrt{3}$ units^2
- 144 units^2
- 6 units^2
- 49 units^2
- 4 units^2
- 81 units^2

M B A R Y C

A S V W O X E

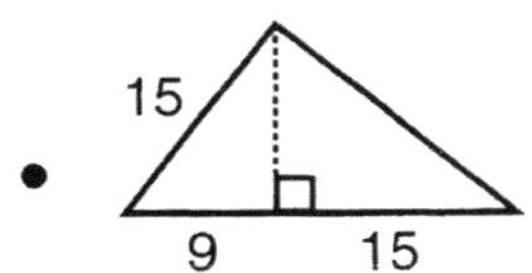

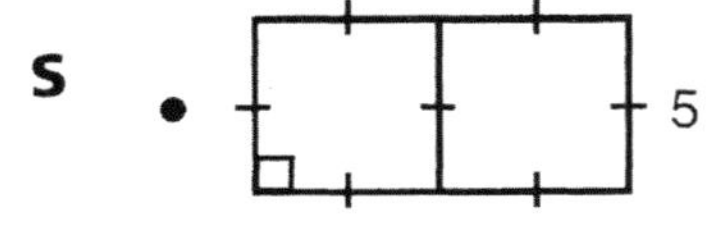

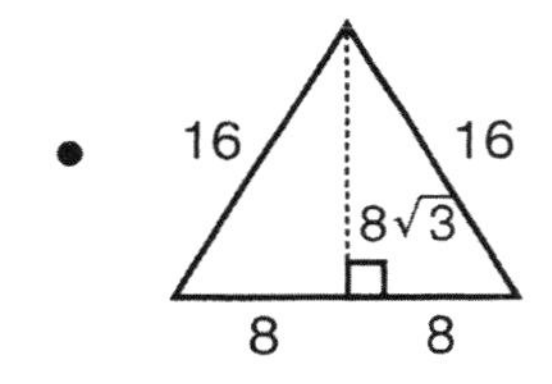

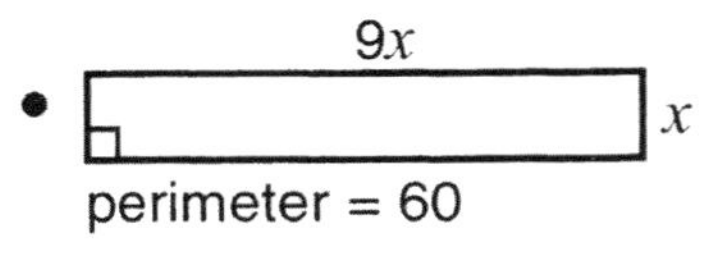

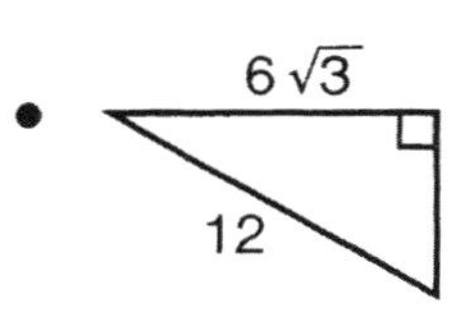

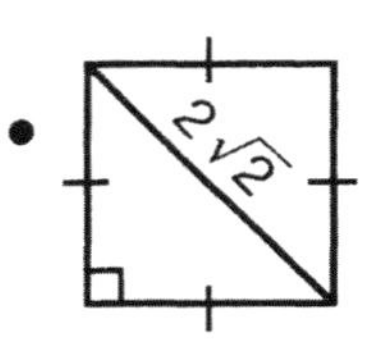

___ ___ ___ ___ ___ !

© Milliken Publishing Company

Name ______________________________

Area of Trapezoids and Parallelograms

Remember

To find the area of a parallelogram or a trapezoid, use these formulas. Or, divide each figure into rectangles and triangles and add the areas together. The height must be perpendicular to the base.

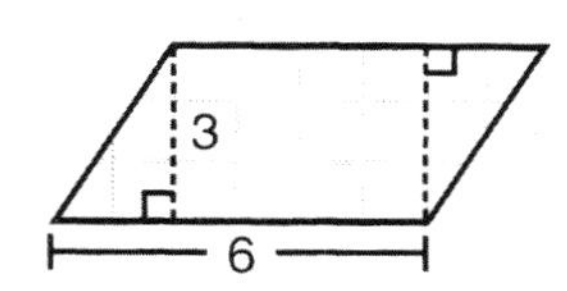

Parallelogram: $A = bh$

Area = base x height

$A = 6 \cdot 3 = 18 \text{ units}^2$

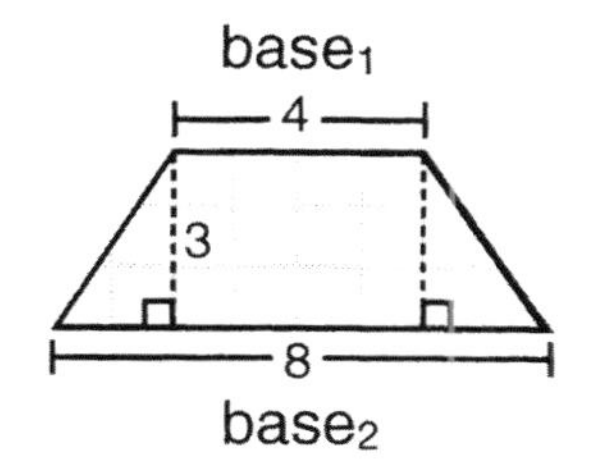

Trapezoid: $A = \frac{1}{2}h(b_1 + b_2)$

Area = $\frac{1}{2}$ x height x (base$_1$ + base$_2$)

$A = \frac{1}{2} \cdot 3 \cdot (4 + 8) = \frac{1}{2} \cdot 3 \cdot 12 = 18 \text{ units}^2$

Find the area of these parallelograms and trapezoids. Then shade in your answers.

1.

24 cm

10 cm

2.

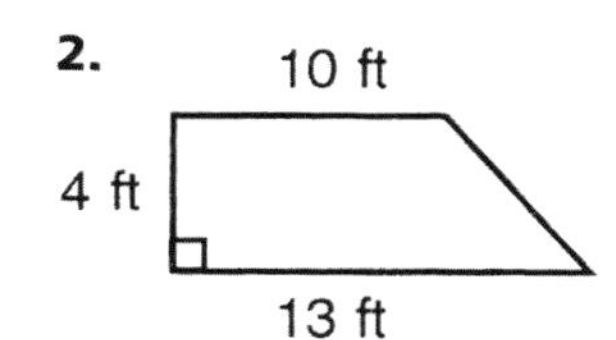

3.

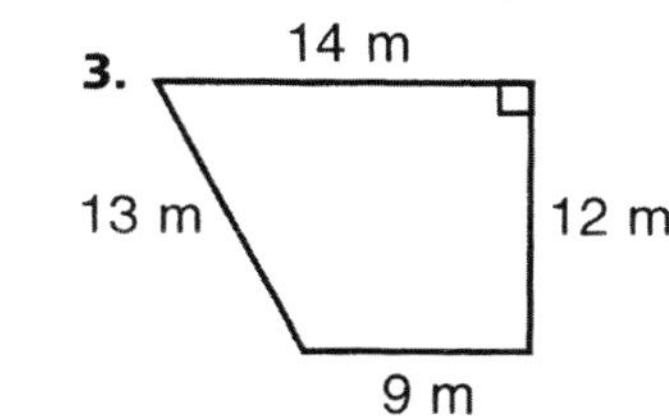

4.

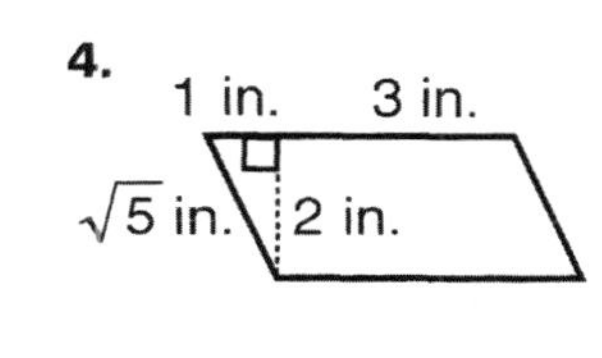

5.

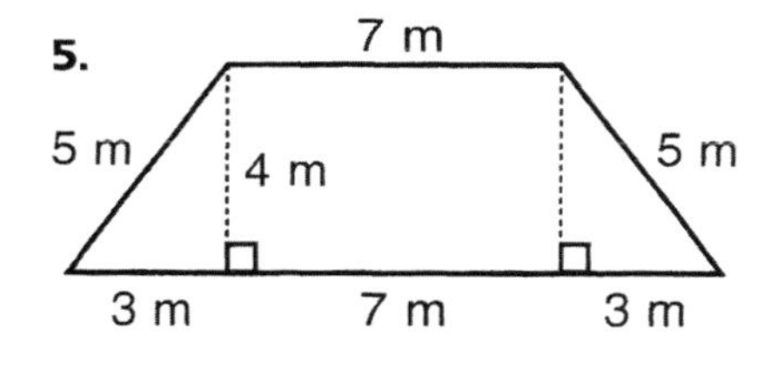

6.

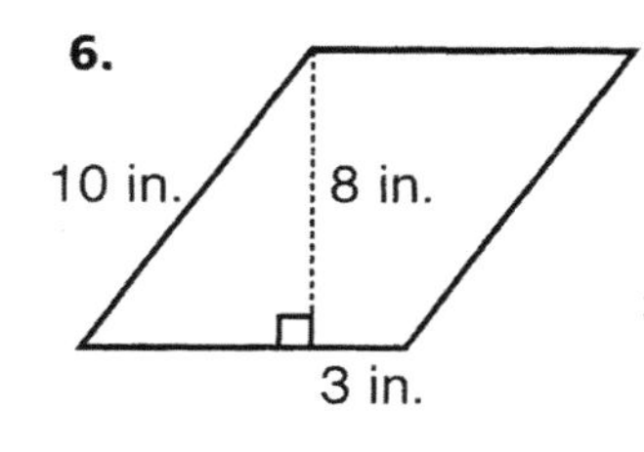

7.

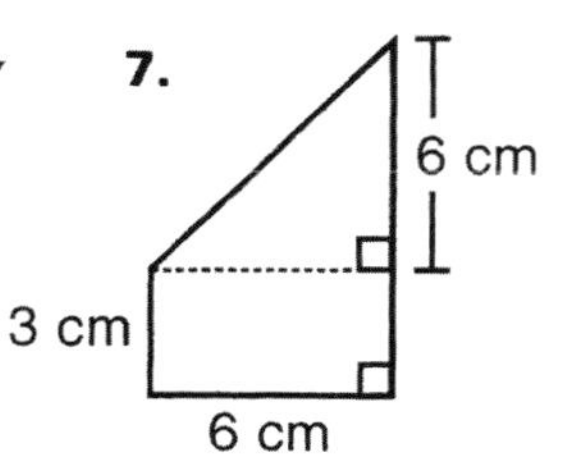

8.

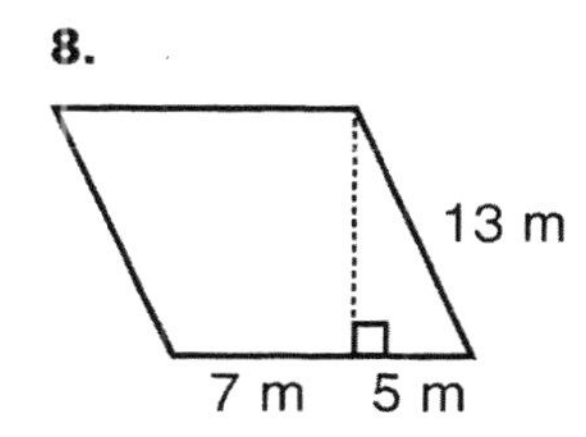

9.

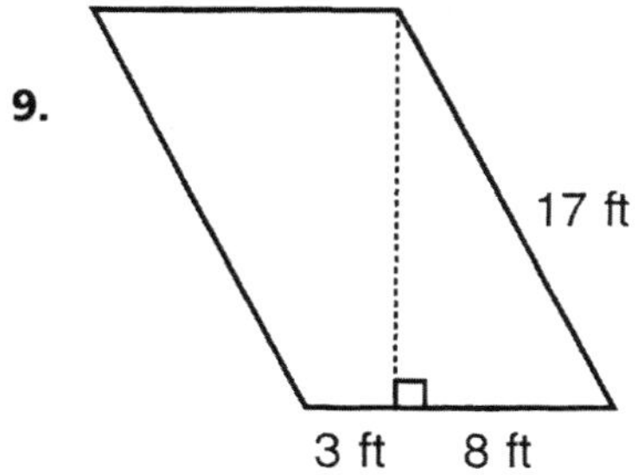

10.

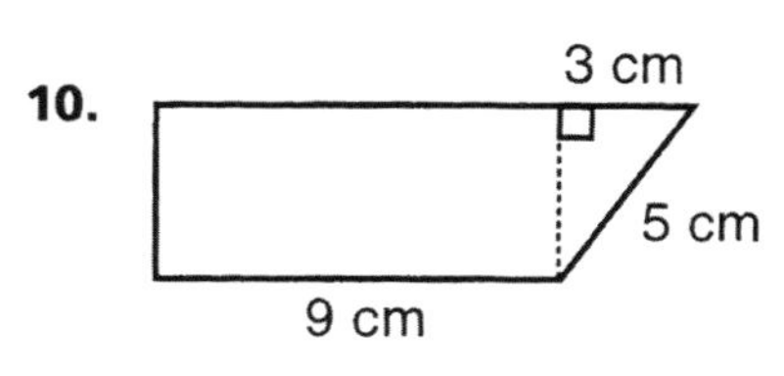

11.

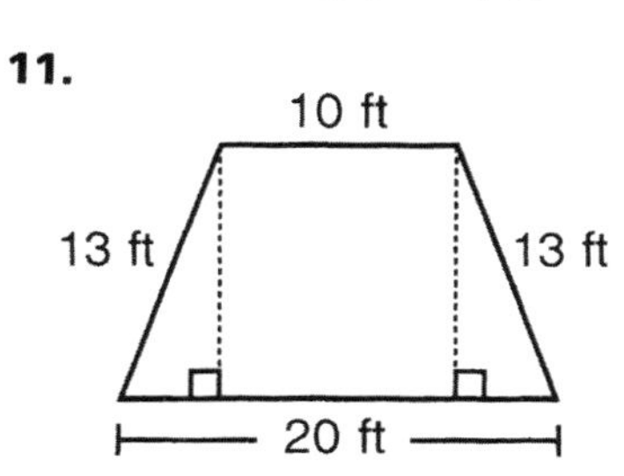

12.

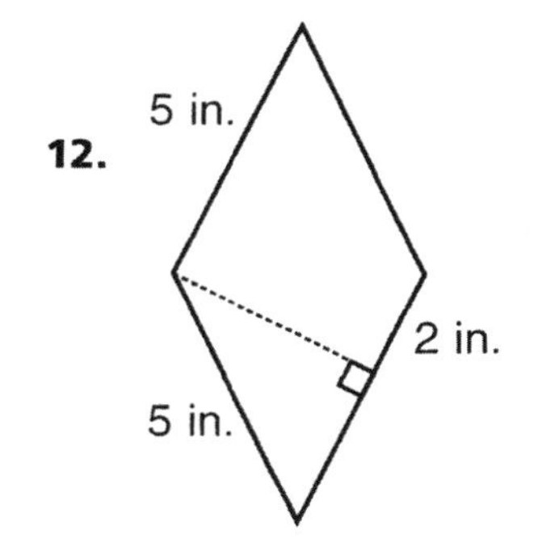

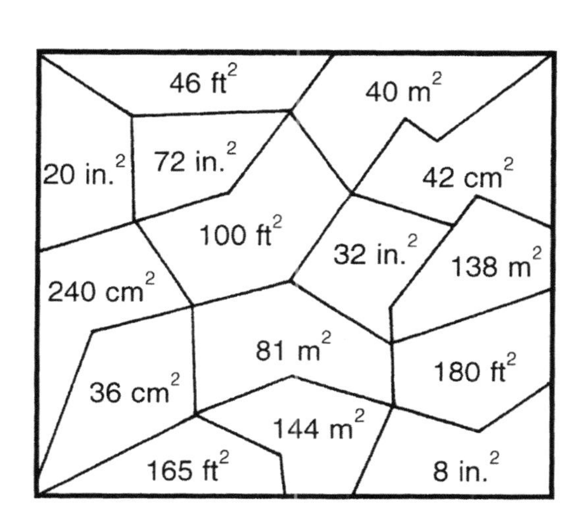

© Milliken Publishing Company

Name ______________________________

Surface Area of Right Prisms

Remember

1. A **right prism** is a solid with two parallel, congruent polygons for bases and rectangles for lateral faces.
2. To find the surface area (SA) of any right prism, add the area of the two bases (2B) and the area of all the lateral faces (LA).

   SA = 2B + LA
3. The surface area formulas for a cube and a right rectangular prism can also be written in these forms:

   $SA = 6s^2$ (s = length of an edge)

   $SA = 2(lw) + 2(lh) + 2(wh)$

Example: Find the surface area of this right triangular prism.

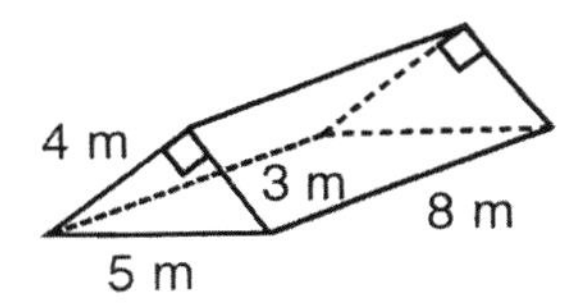

Area of triangular base $= \frac{1}{2} \cdot 3 \cdot 4 = 6$

Area of lateral faces $= 3 \cdot 8 = 24$

$4 \cdot 8 = 32$

$5 \cdot 8 = 40$

$SA = 2B + LA$

$= 2(6) + (24 + 32 + 40)$

$= 12 + 96 = 108 \text{ m}^2$

Find the surface area of these right prisms. Place your answers in the cross-number puzzle.

m²

Across

2. A cube with an edge length of 6 m.

4.

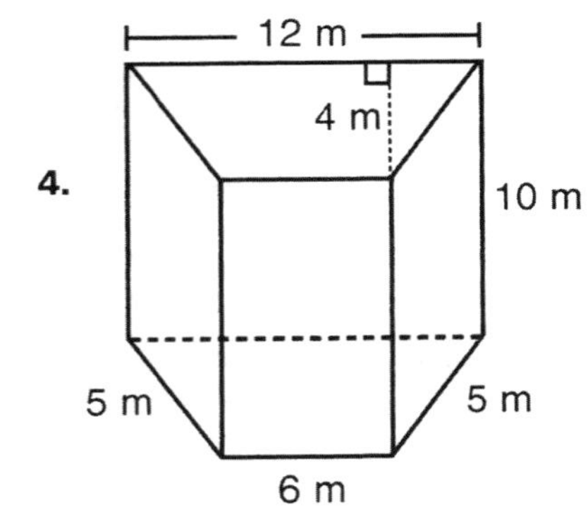

5.

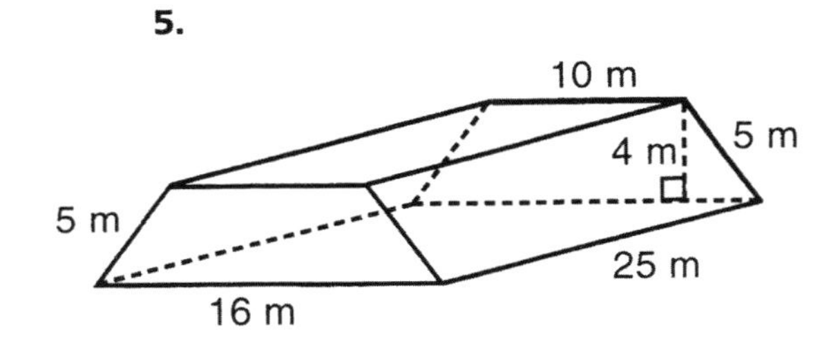

7.

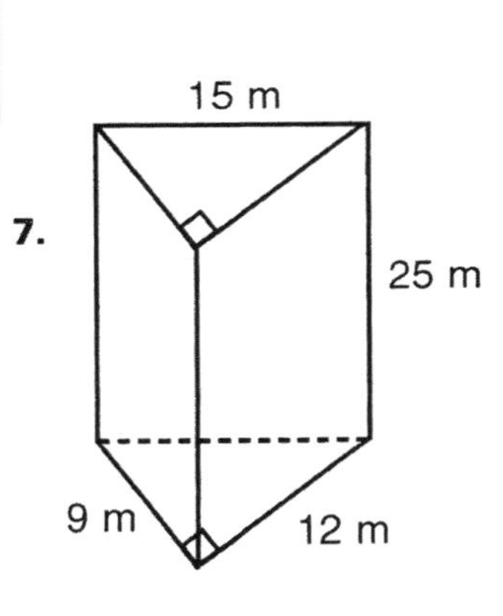

Down

1. A right rectangular prism with a length of 12 m, a width of 5 m, and a height of 8 m.

3.

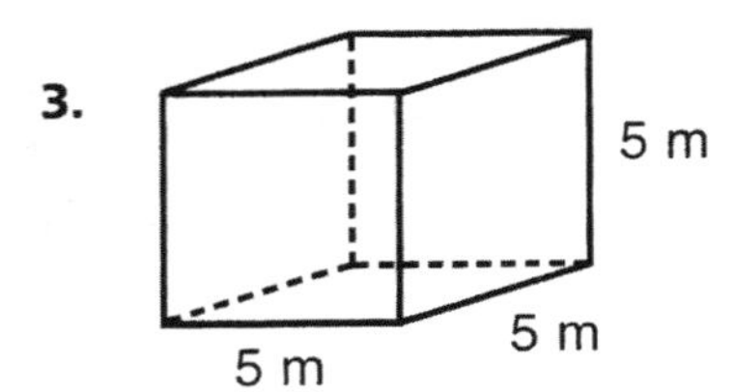

4.

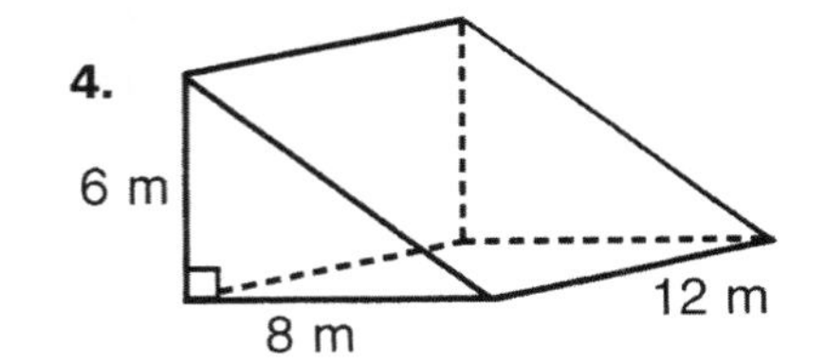

6.

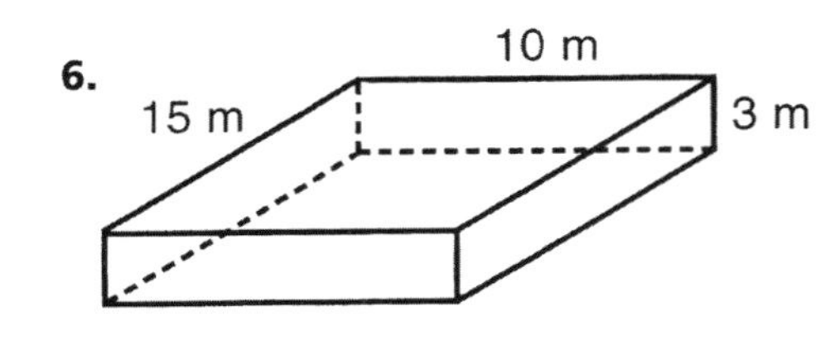

© Milliken Publishing Company

Name ______________________________

Volume of Right Prisms

Remember

1. **Volume** is the space inside a solid. It is measured in cubic units.
2. To find the volume of any right prism, multiply the area of one base (B) by the height (h) of the prism.

   $V = Bh$ Volume = (area of base) x height
3. The volume formulas for a cube and a right rectangular prism can also be written in these forms:

   $V = s^3$

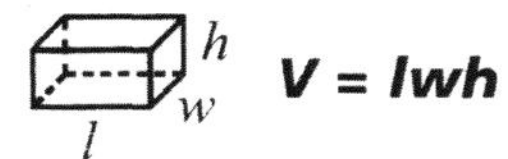

   $V = lwh$

Example: Find the volume of this right triangular prism.

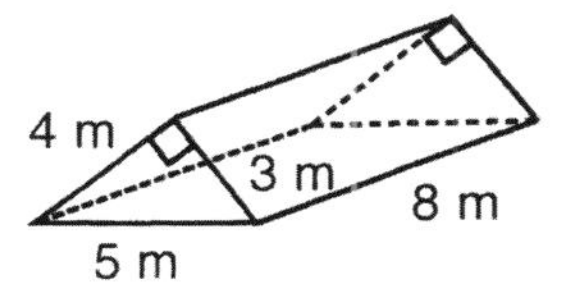

Area of triangular base = $\frac{1}{2} \cdot 3 \cdot 4 = 6$

$V = Bh$

$= 6 \cdot 8 = 48 \text{ m}^3$

Find the volume of these right prisms. Use the answer code to reveal the name of a famous Persian mathematician, astronomer, and poet who lived about 1,000 years ago.

1.

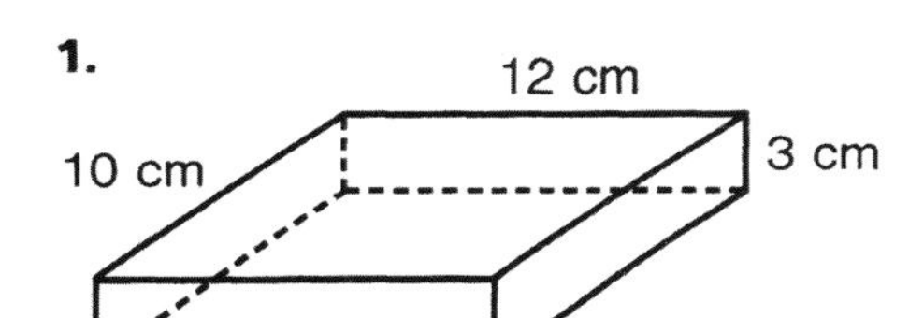

2.

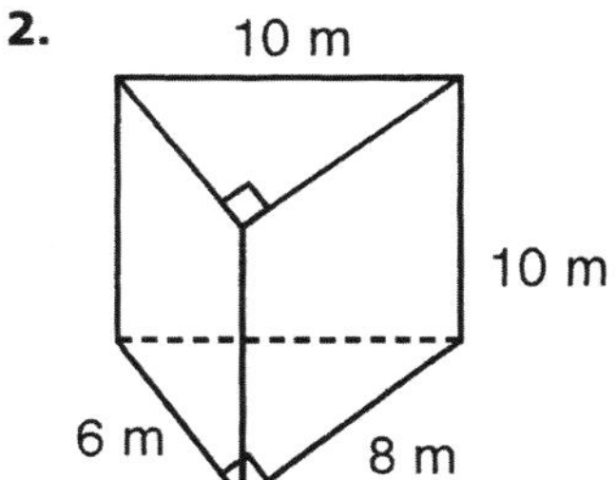

3.

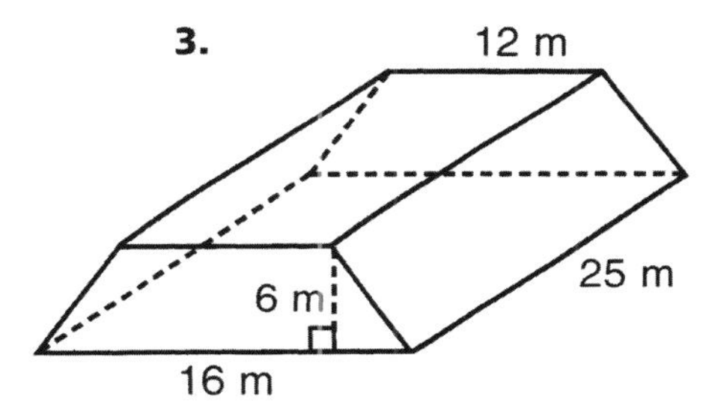

4.

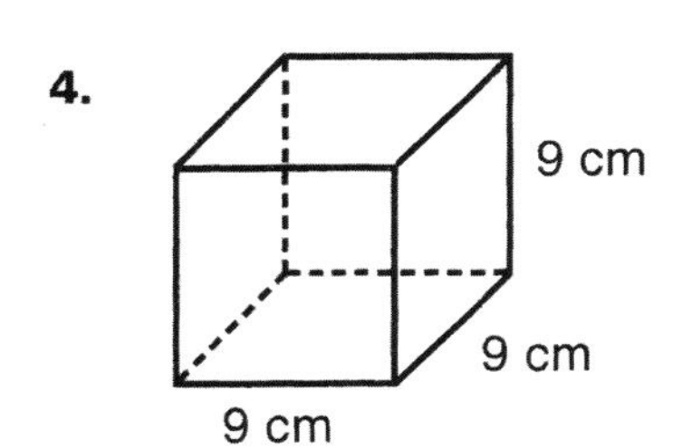

5.

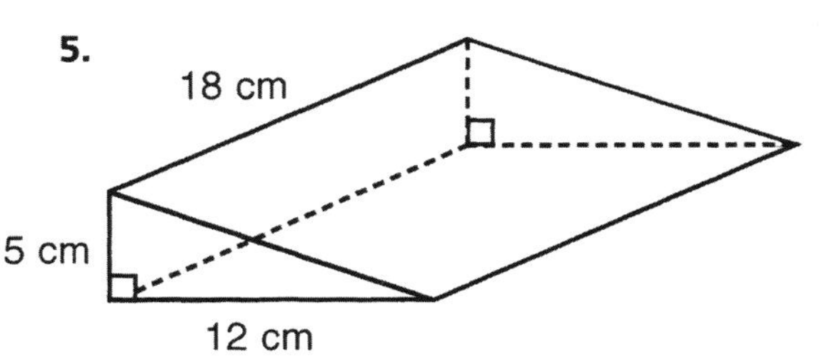

6.

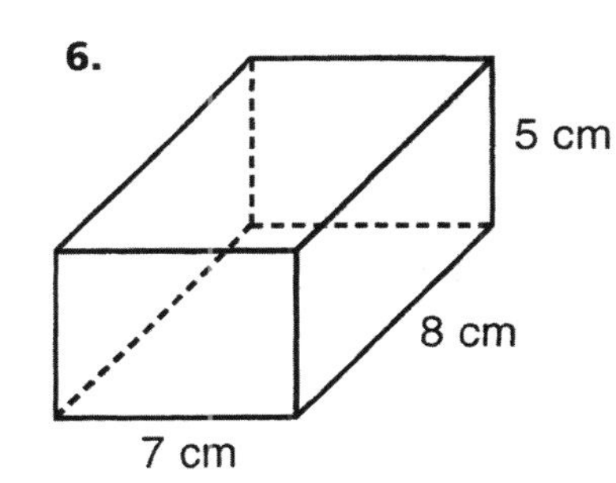

7.

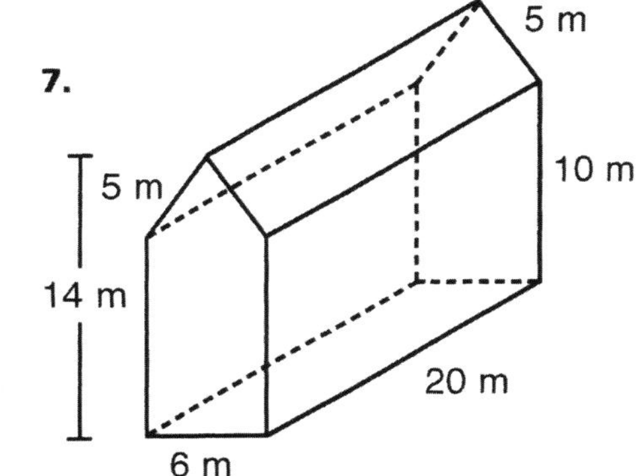

240 m^3	280 cm^3	360 cm^3	540 cm^3	729 cm^3	1440 m^3	2100 m^3
A	H	K	M	O	R	Y

__ __ __ __   __ __ __ __ __ __ __

4 5 2 7   1 6 2 3 3 2 5

© Milliken Publishing Company

Name ______________________________

Circumference and Area of Circles

Remember

1. **Circumference** is the distance around a circle. Think of it as the circle's perimeter.

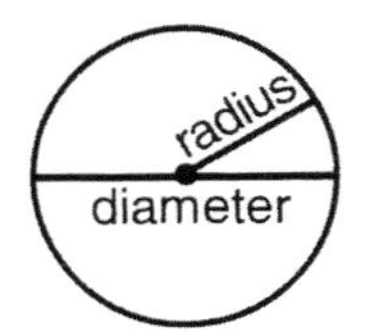

Circumference = π x diameter
$C = \pi d$ or
$C = 2\pi r$ (The diameter is twice the length of the radius: $d = 2r$.)

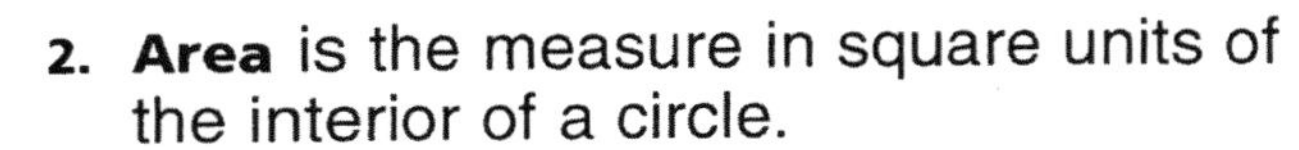

2. **Area** is the measure in square units of the interior of a circle.

Area = π x radius x radius $A = \pi r^2$

Example: Find the circumference and area of this circle. Use 3.14 as an approximation for π.

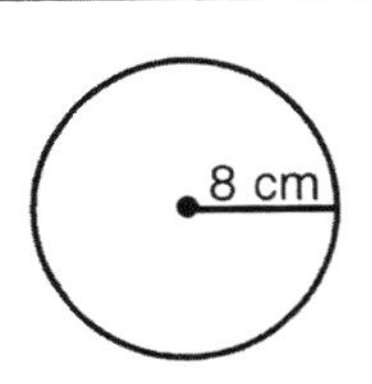

$C = 2\pi r$
$\approx 2 \cdot 3.14 \cdot 8$
≈ 50.24
≈ 50 cm

$A = \pi r^2$
$\approx 3.14 \cdot 8^2$
≈ 200.96
≈ 201 cm^2

Draw straight lines to match each radius of a circle to its correct circumference and area. Use 3.14 for π. Write the uncrossed letters in the empty circles below to answer the riddle.

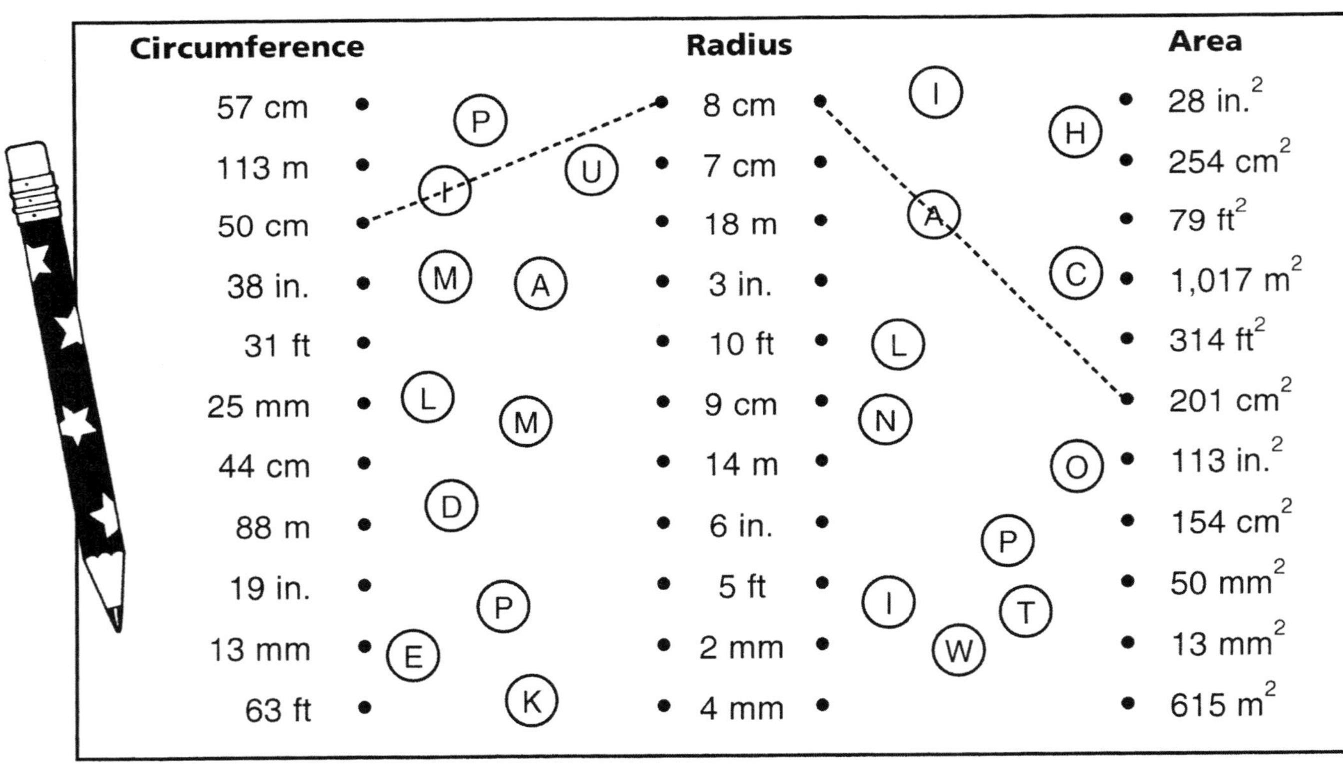

What do you get when you take the circumference of a jack-o'-lantern and divide it by its diameter?

Name ______________________

Area of a Shaded Region

Example

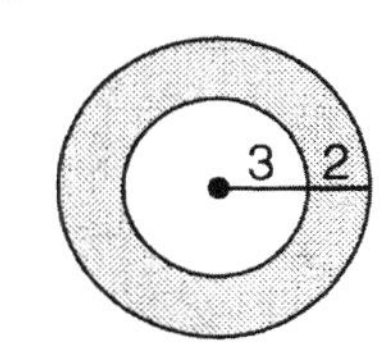

Find the area of the shaded region. Leave the answer in terms of π.

1. Analyze the problem.
Area of the shaded ring = (Area of shaded circle) – (Area of circle)

2. Review the needed area formula(s).
Area of a circle = πr^2

3. Solve.
Shaded area = $5^2\pi - 3^2\pi = 25\pi - 9\pi = 16\pi$ units^2

Match each diagram to its solution equation. Then find the area of each shaded region. Leave the answers in terms of π. Find and circle each answer in the box below.

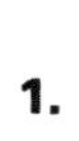

1. 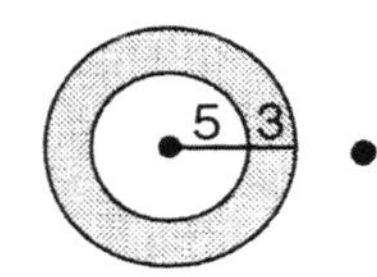

2. 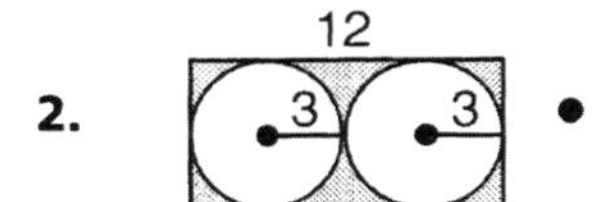

3. 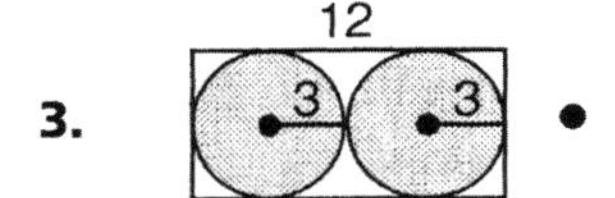

4. 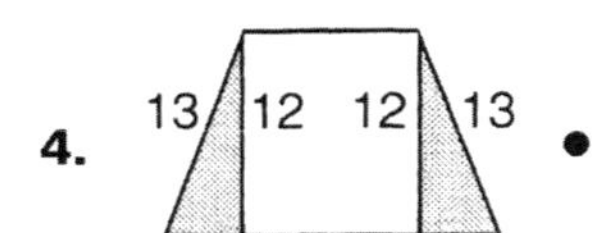

5. 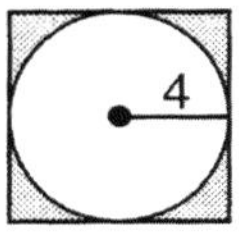 (diagram: 4)

6. 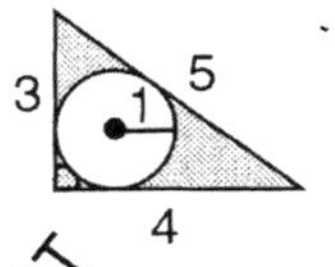 (diagram: 3, 5, 1, 4)

7. 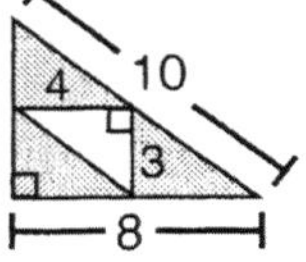 (diagram: 4, 10, 3, 8)

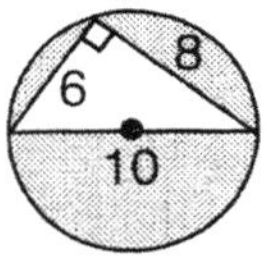

8. (diagram: 6, 8, 10)

- (A of square) – (A of circle) = __________ units^2
- (A of shaded circle) – (A of circle) = __________ units^2
- (A of shaded circle) + (A of shaded circle) = __________ units^2
- (A of rectangle) – (A of circle + A of circle) = __________ units^2
- (A of shaded circle) – (A of triangle) = __________ units^2
- (A of triangle) + (A of triangle) = __________ units^2
- (A of shaded triangle) – (A of circle) = __________ units^2
- (A of shaded triangle) – (A of triangle) = __________ units^2

$6 - \pi$	18	18π	$25\pi - 24$	39π	60	$64 - 16\pi$	$72 - 18\pi$

© Milliken Publishing Company

Name ______________________________

Surface Area and Volume of Cylinders

Remember

1. A **right cylinder** is a solid with two parallel, congruent circles for bases. Its lateral surface is curved and is perpendicular to the bases.
2. To find the surface area (SA) of a right cylinder, add the area of the two bases (2B) and the lateral area (LA). The lateral area is equal to the product of the circumference and the height.

   SA = 2B + LA
3. To find the volume of a right cylinder, multiply the area of one base (B) by the height (h) of the cylinder.

   V = Bh Volume = (area of base) x height

Example:
Find the surface area and volume of this right cylinder.

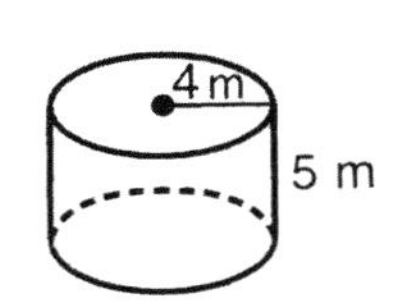

Area of circular base $= \pi \cdot 4^2 = 16\pi$

Lateral area = circumference x height

$= (2\pi r)h$

$= 2 \cdot \pi \cdot 4 \cdot 5 = 40\pi$

$SA = 2B + LA$

$= 2(16\pi) + (40\pi)$

$= 32\pi + 40\pi = 72\pi \text{ m}^2 \approx 226 \text{ m}^2$

$V = Bh$

$= (16\pi)5 = 80\pi \text{ m}^3 \approx 251 \text{ m}^3$

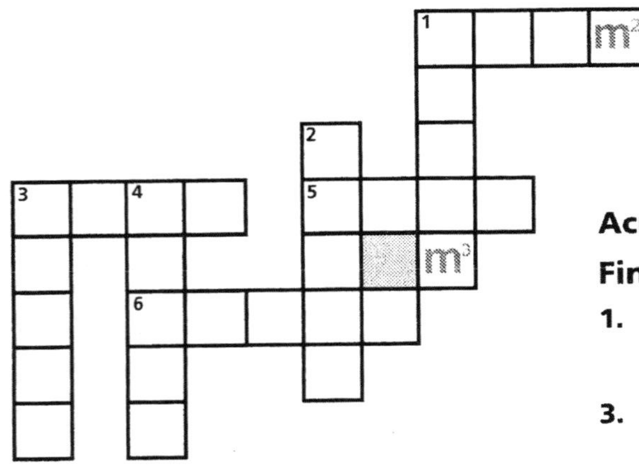

Find the surface area or volume of these right cylinders. Use 3.14 for π, and round your answers to the nearest whole number. Place your answers in the cross-number puzzle.

Across

Find the surface area:

1. Right cylinder with a radius of 4 m and a height of 6 m.
3. Right cylinder with a diameter of 10 m and a height of 5 m.
5. Right cylinder with a radius of 3 m and a height of 10 m.
6. Right cylinder with a diameter of 20 m and a height of 12 m.

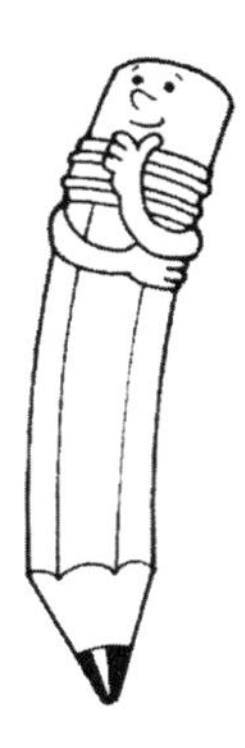

Down

Find the volume:

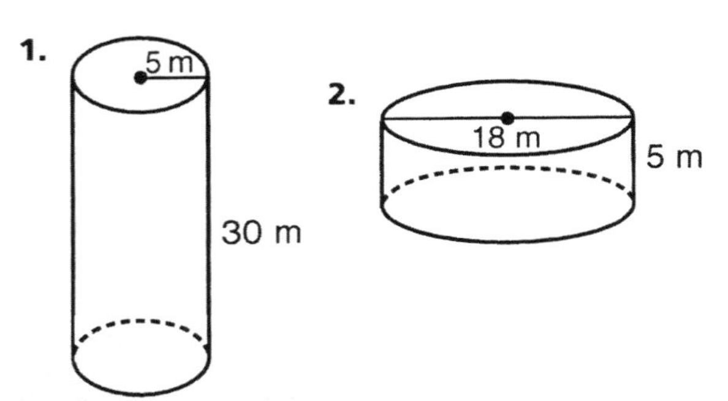

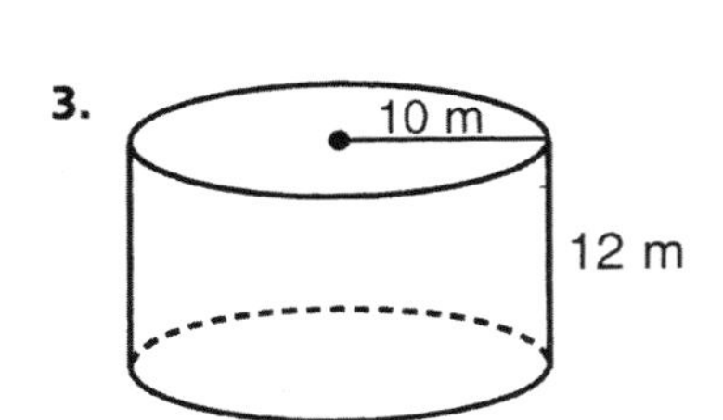

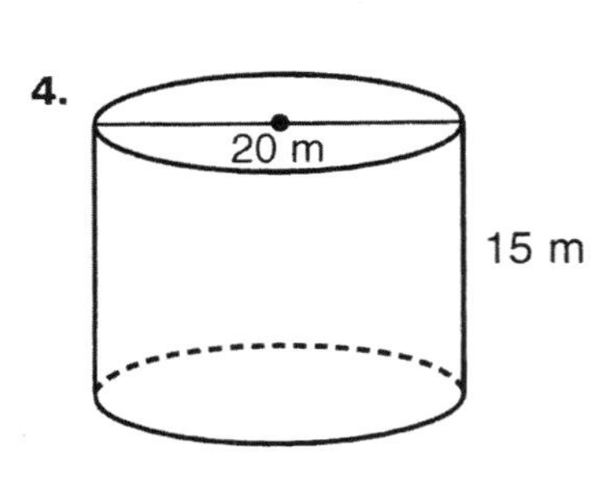

© Milliken Publishing Company

Name ______________________________

Arcs, Central Angles, and Inscribed Angles

Remember

1. An angle whose vertex is the center of a circle is a **central angle**. Example: $\angle BPC$
2. An **arc** is a curve of a circle. It is named by its endpoints.

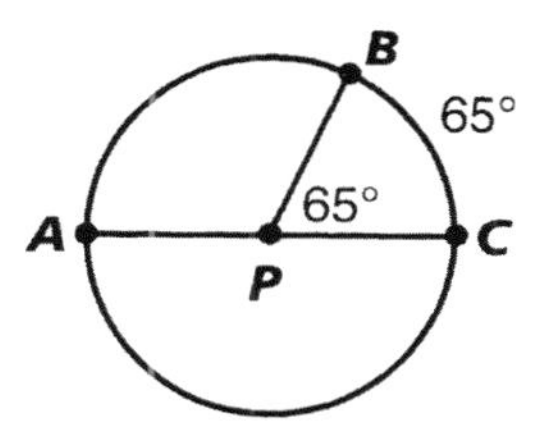

A whole circle measures 360°.

A **minor arc** measures less than 180°. Its measure is equal to the measure of its central angle. Example: $m\widehat{BC} = m\angle BPC = 65°$

A **semicircle** measures 180°. Its central angle is a diameter. Example: $m\widehat{AC} = m\angle APC = 180°$

A **major arc** measures more than 180°. Its measure is the difference between 360° and the measure of its central angle. Example: $m\widehat{BAC} = 360° - 65° = 295°$

3. A **chord** is a segment whose endpoints are points on a circle. Example: $\overline{DF}$
An **inscribed angle** is an angle whose sides are chords and whose vertex is a point on the circle. Example: $\angle DFE$

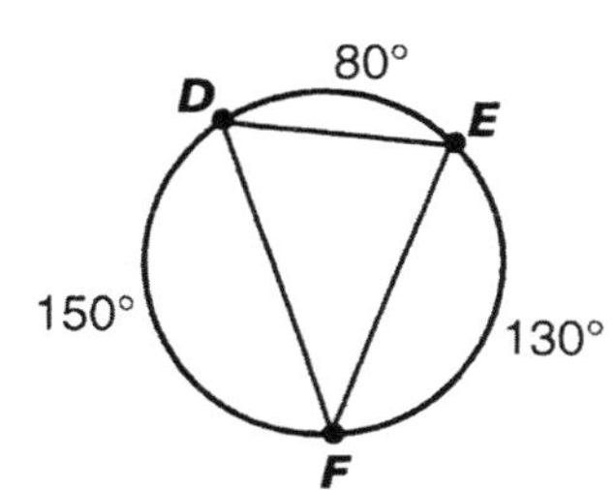

4. The measure of an inscribed angle is equal to half the measure of its intercepted arc.

$m\angle DFE = \frac{1}{2}\ m\widehat{DE} = \frac{1}{2}\ (80°) = 40°$

$m\angle EDF = \frac{1}{2}\ m\widehat{EF} = \frac{1}{2}\ (130°) = 65°$

$m\angle DEF = \frac{1}{2}\ m\widehat{DF} = \frac{1}{2}\ (150°) = 75°$

Note: The sum of the angles of $\triangle DEF$ = 180°.

Draw lines to match each arc or angle to its measure. Solve for the missing measures.

1.	$m\widehat{XY}$ •	• 180°
2.	$m\angle XOZ$ •	• ____
3.	$m\angle YOZ$ •	• 50°
4.	$m\widehat{XZY}$ •	• 230°
5.	$m\widehat{YXZ}$ •	• 130°

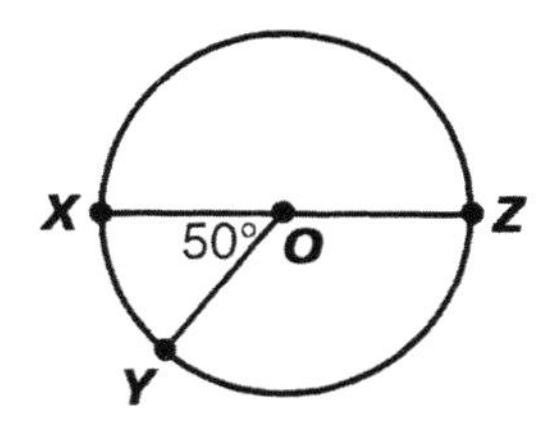

O is the center point.
$\overline{XZ}$ is a diameter.

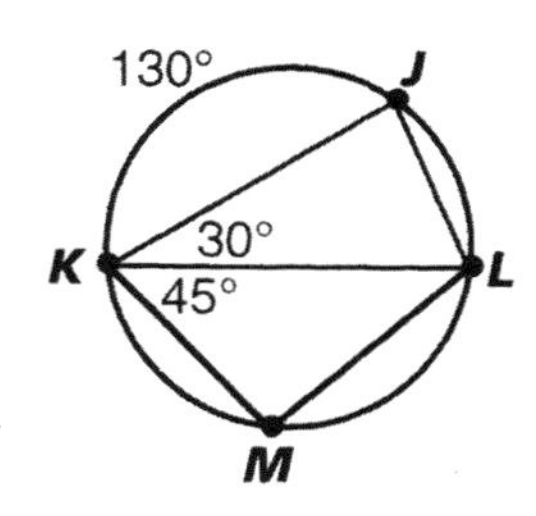

6.	$m\widehat{JL}$ •	• ____
7.	$m\angle KLJ$ •	• 95°
8.	$m\angle KJL$ •	• 85°
9.	$m\widehat{LM}$ •	• 60°
10.	$m\angle KML$ •	• 65°

11.	$m\widehat{BC}$ •	• 100°
12.	$m\angle BEC$ •	• 80°
13.	$m\widehat{BEC}$ •	• 40°
14.	$m\widehat{DC}$ •	• ____
15.	$m\angle EBD$ •	• 280°

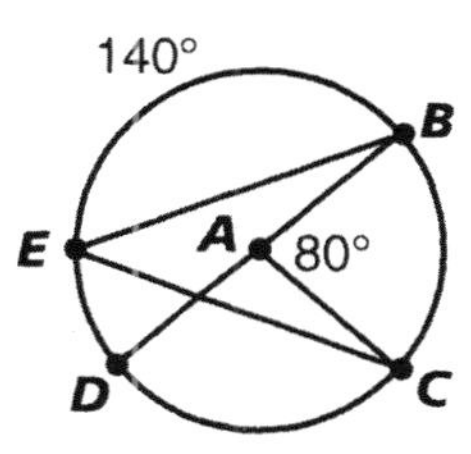

A is the center point.
$\overline{DB}$ is a diameter.

© Milliken Publishing Company

Name ______________________________

Angles Formed by Chords, Secants, and Tangents

Remember

1. A **chord** is a segment whose endpoints are points on a circle.
   A **secant** is a line that intersects a circle at two points.
   A **tangent** is a line that intersects a circle at exactly one point.

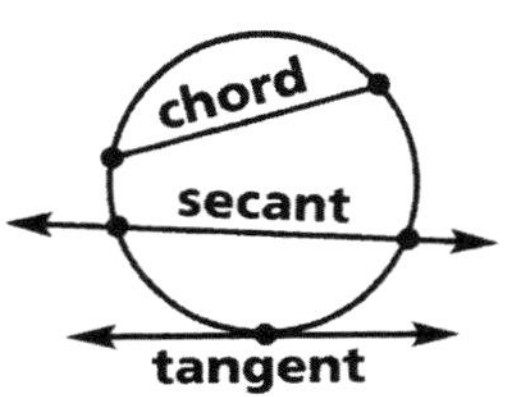

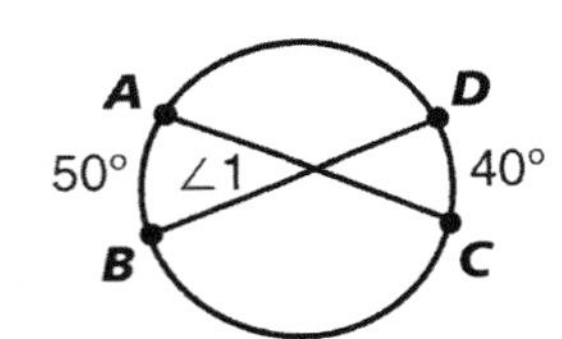

2. If two chords intersect *inside* a circle, the measure of each angle is equal to half the measure of the *sum* of the intercepted arcs.

$$m\angle 1 = \tfrac{1}{2}\,(m\overset{\frown}{AB} + m\overset{\frown}{CD}) = \tfrac{1}{2}(50^\circ + 40^\circ) = \tfrac{1}{2}\,(90^\circ) = 45^\circ$$

3. If two secants, a secant and a tangent, or two tangents intersect *outside* a circle, the measure of the angle formed is equal to half the measure of the *difference* of the intercepted arcs.

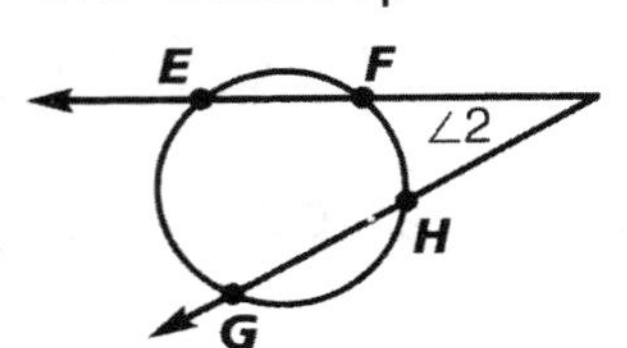

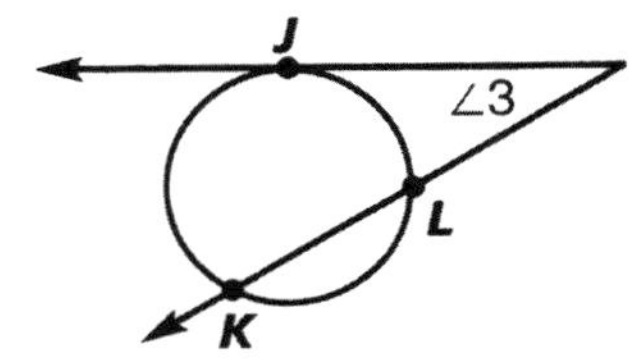

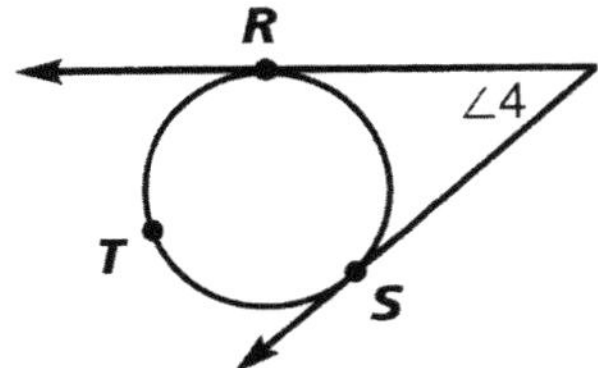

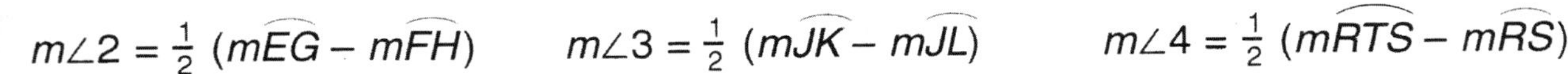

$m\angle 2 = \tfrac{1}{2}\,(m\overset{\frown}{EG} - m\overset{\frown}{FH})$ $m\angle 3 = \tfrac{1}{2}\,(m\overset{\frown}{JK} - m\overset{\frown}{JL})$ $m\angle 4 = \tfrac{1}{2}\,(m\overset{\frown}{RTS} - m\overset{\frown}{RS})$

Use the diagrams above and the information given to find the missing measures. Then use the decoder to answer this riddle: *Why did the polite man spend the day at the beach?*

1. $m\overset{\frown}{AB} = 62^\circ$, $m\overset{\frown}{CD} = 88^\circ$, $m\angle 1 =$ _____
2. $m\overset{\frown}{AB} = 25^\circ$, $m\angle 1 = 40^\circ$, $m\overset{\frown}{CD} =$ _____
3. $m\overset{\frown}{AB} = 50^\circ$, $m\overset{\frown}{CD} = 30^\circ$, $m\angle 1 =$ _____
4. $m\overset{\frown}{EG} = 75^\circ$, $m\overset{\frown}{FH} = 25^\circ$, $m\angle 2 =$ _____
5. $m\overset{\frown}{FH} = 40^\circ$, $m\overset{\frown}{EG} = 100^\circ$, $m\angle 2 =$ _____
6. $m\angle 2 = 20^\circ$, $m\overset{\frown}{FH} = 75^\circ$, $m\overset{\frown}{EG} =$ _____
7. $m\overset{\frown}{JK} = 145^\circ$, $m\overset{\frown}{JL} = 45^\circ$, $m\angle 3 =$ _____
8. $m\angle 3 = 45^\circ$, $m\overset{\frown}{JK} = 180^\circ$, $m\overset{\frown}{JL} =$ _____
9. $m\overset{\frown}{KL} = 100^\circ$, $m\overset{\frown}{JK} = 210^\circ$, $m\angle 3 =$ _____
10. $m\overset{\frown}{RTS} = 275^\circ$, $m\angle 4 =$ _____
11. $m\overset{\frown}{RS} = 115^\circ$, $m\angle 4 =$ _____
12. $m\overset{\frown}{TR} = 120^\circ$, $m\overset{\frown}{TS} = 130^\circ$, $m\angle 4 =$ _____

25°	30°	40°	50°	55°	65°	70°	75°	80°	90°	95°	115°
A	B	C	D	E	G	H	N	S	T	U	W

__ __ __ __ __ __ __   __ __   __ __ __ __ __ __ TO BE A __ __ __   __ __ __ __ !

5 2 3 4 10 9 2   12 2   6 4 1 8 2 7   8 4 1   11 2 1 8

© Milliken Publishing Company

Name ______________________________

Length of Chords

Remember

If two chords intersect, the product of the segment lengths along one chord is equal to the product of the segment lengths along the other chord.

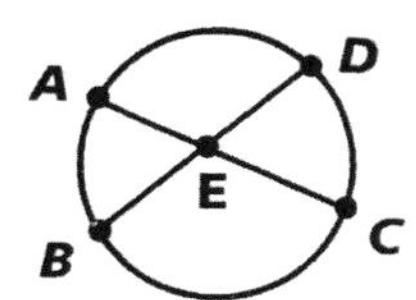

$AE \cdot EC = BE \cdot ED$

Examples: Find the value of x and the length of $\overline{AC}$.

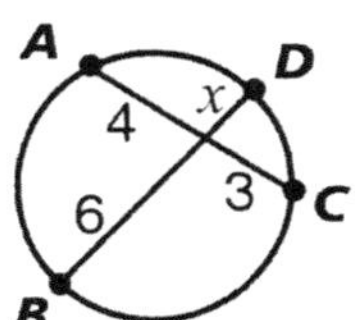

$4 \cdot 3 = 6 \cdot x$
$12 = 6x$
$2 = x$
$AC = 4 + 3 = 7$

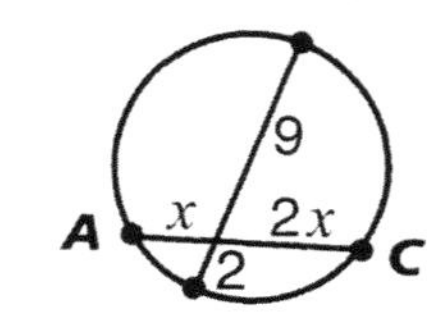

$x \cdot 2x = 2 \cdot 9$
$2x^2 = 18$
$x^2 = 9$
$x = 3$

Check: $3 \cdot 2(3) \stackrel{?}{=} 2 \cdot 9$
$3 \cdot 6 \stackrel{?}{=} 2 \cdot 9$
$18 = 18$
$AC = x + 2x = 3 + 6 = 9$

Find the value of x and the length of $\overline{AC}$. Follow your answers in order through the maze.

1.

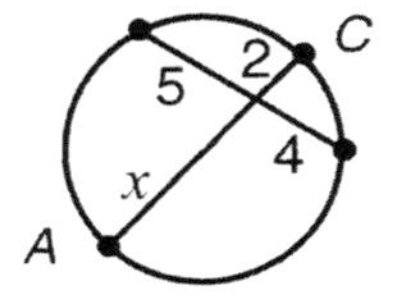

$x =$ _____ $AC =$ _____

2.

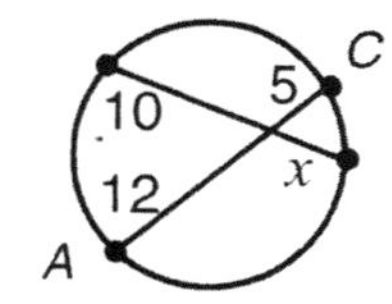

$x =$ _____ $AC =$ _____

3.

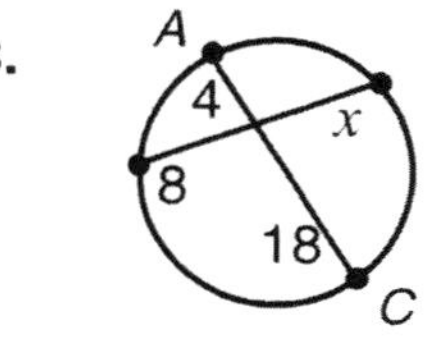

$x =$ _____ $AC =$ _____

4.

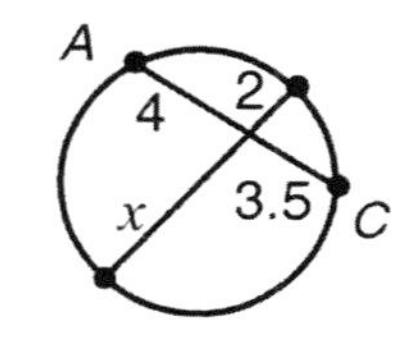

$x =$ _____ $AC =$ _____

5.

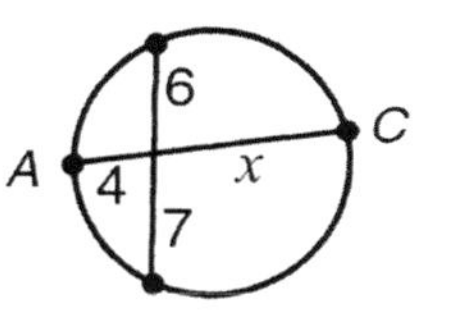

$x =$ _____ $AC =$ _____

6.

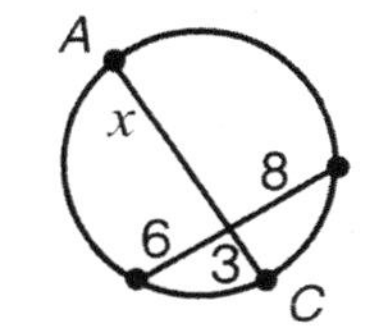

$x =$ _____ $AC =$ _____

7.

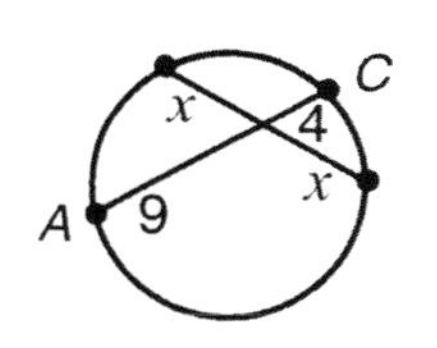

$x =$ _____ $AC =$ _____

8.

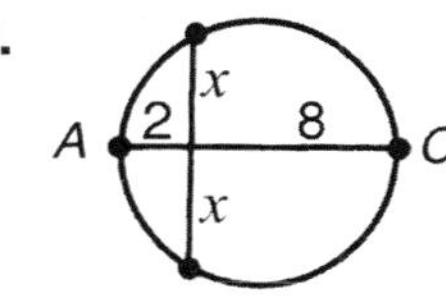

$x =$ _____ $AC =$ _____

9.

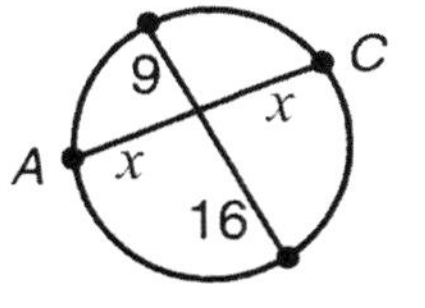

$x =$ _____ $AC =$ _____

10.

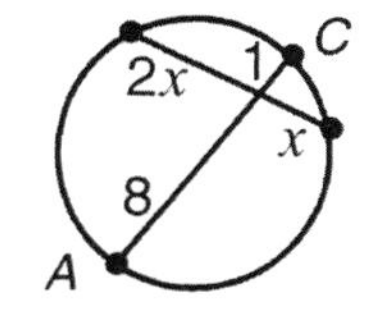

$x =$ _____ $AC =$ _____

11.

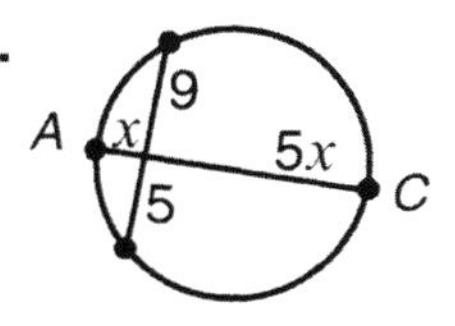

$x =$ _____ $AC =$ _____

12.

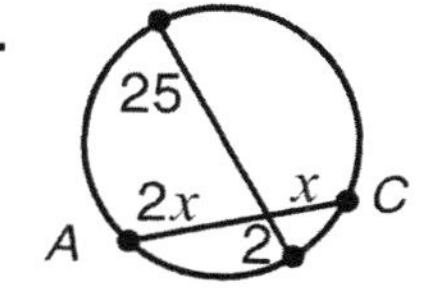

$x =$ _____ $AC =$ _____

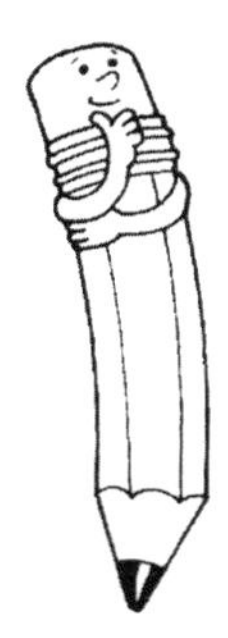

START

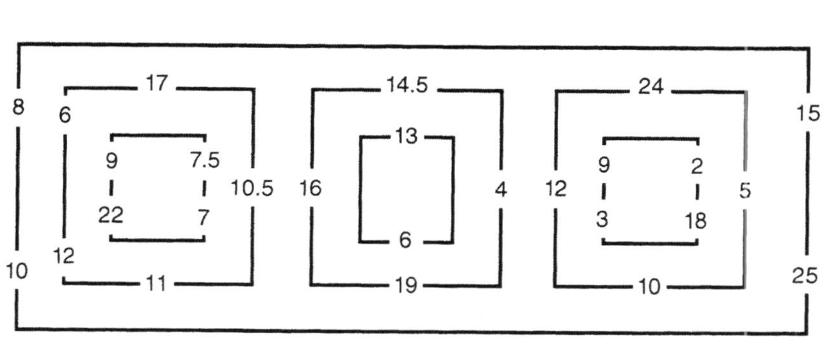

FINISH

© Milliken Publishing Company

Name ______________________________

Length of Secant and Tangent Segments

Remember

1. If two secant segments share the same point outside a circle, the product of the length of one secant and its external segment length is equal to the product of the length of the other secant and its external segment length.

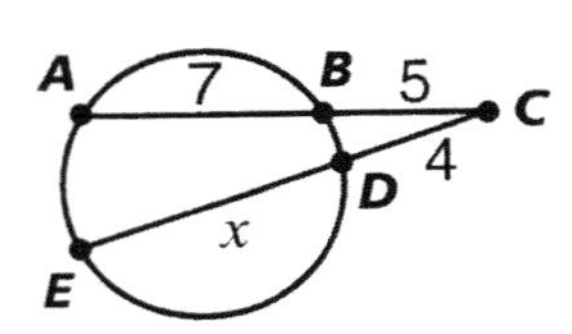

$AC \cdot BC = EC \cdot DC$

Example:
Given: $AB = 7$, $BC = 5$, $DC = 4$. Find the length of $\overline{ED}$ and $\overline{EC}$.

$$AC \cdot BC = EC \cdot DC$$
$$(7 + 5) \cdot 5 = (x + 4) \cdot 4$$
$$60 = 4x + 16$$
$$44 = 4x$$
$$11 = x$$

$ED = 11$; $EC = 15$

2. If a secant segment and a tangent segment share the same external point, the segment lengths follow a similar product rule.

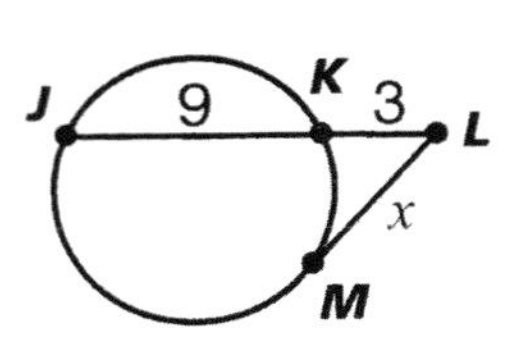

$JL \cdot KL = ML \cdot ML$

Example:
Given: $JK = 9$, $KL = 3$. Find the length of $\overline{ML}$.

$$JL \cdot KL = ML \cdot ML$$
$$(9 + 3) \cdot 3 = x \cdot x$$
$$36 = x^2$$
$$6 = x$$

$ML = 6$

3. If two tangent segments share the same external point, they are congruent.

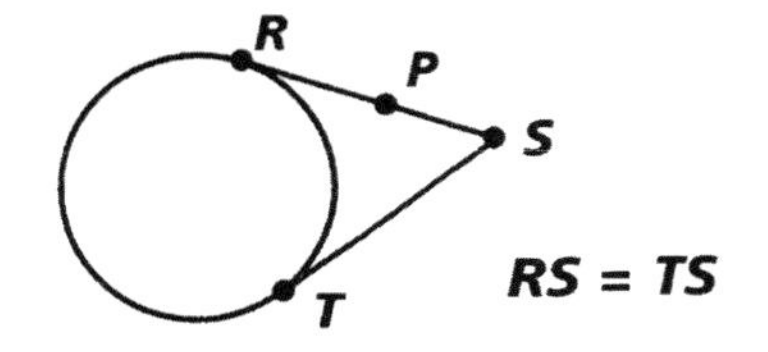

$RS = TS$

Use the diagrams above and the measures given to find the missing lengths.

1. $AC = 15$, $BC = 5$, $DC = 3$, $ED =$ _____, $EC =$ _____, $AB =$ _____
2. $ED = 20$, $DC = 4$, $BC = 8$, $EC =$ _____, $AB =$ _____, $AC =$ _____
3. $JK = 30$, $KL = 10$, $JL =$ _____, $ML =$ _____
4. $ML = 12$, $KL = 10$, $JK =$ _____, $JL =$ _____
5. $JL = 25$, $ML = 20$, $KL =$ _____, $JK =$ _____
6. $RP = 10$, $TS = 17$, $RS =$ _____, $PS =$ _____

4	A
4.4	C
7	E
9	F
10	G
12	H
14.4	I
16	M
17	N
20	O
22	R
24	S
25	T
40	W

Use the decoder to find the name of an Italian mathematician and the math curve named for her. The Italian word for curve was mistranslated which is how the curve got its odd name.

M _____ _____ _____ _____   _____ _____ _____ _____ _____ _____ ;
5-KL 2-AB 1-ED 4-JL 2-AB   2-AB 1-AB 6-RS 6-PS 2-EC 4-JL

_____ _____ _____ _____ _____   _____ _____   _____ _____ _____ _____ _____ _____
3-JL 4-JL 1-EC 4-JK 2-AC   3-ML 5-JK   2-AB 1-AB 6-RS 6-PS 2-EC 4-JL

© Milliken Publishing Company

Name ____

Area of a Sector and Arc Length

Remember

1. A **sector** is a section of a circle that is bounded by two radii and their intercepted arc. The area of a sector is the area of a fraction of the entire circle.

   To find the fraction, divide the measure of the arc (m) by the measure of the entire circle (360°). Multiply that fraction by the area of a circle (πr^2) to find the area of the sector.

   Area of a sector $= \frac{m}{360°} \cdot \pi r^2$

B 60° A
60°
12 cm
P

Example: Find the area of sector BPA.

$$A = \frac{60°}{360°} \cdot \pi \cdot (12 \text{ cm})^2$$
$$= \frac{1}{6} \cdot 144\pi \text{ cm}^2$$
$$= 24\pi \text{ cm}^2$$

2. **Arc length** is a fraction of the circumference of a circle. Multiply the fraction of the circle by the circumference ($2\pi r$) to determine the arc length.

   Arc length $= \frac{m}{360°} \cdot 2\pi r$

Example: Find the length of $\overset{\frown}{XY}$.

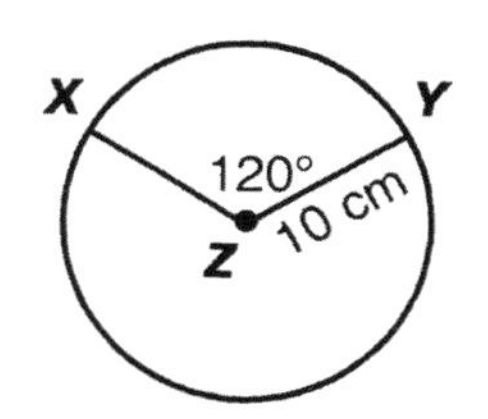

$$\textit{Arc length} = \frac{120°}{360°} \cdot 2\pi(10 \text{ cm})$$
$$= \frac{1}{3} \cdot 20\pi \text{ cm}$$
$$= \frac{20}{3}\pi \text{ cm}$$

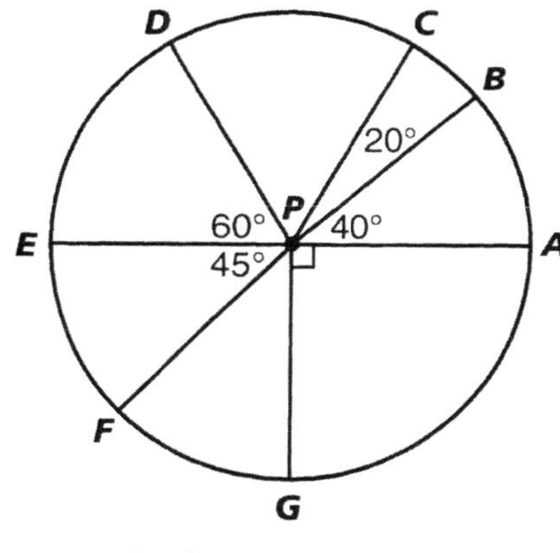

P is the center point.
$\overline{EA}$ is a diameter.

Use the diagram and each given radius or diameter to find the area of the sector (in square meters) or the length of the arc named (in meters). Then shade in your answers.

1. EP = 6 m; Area of sector EPD = ____________
2. DP = 9 m; Area of sector EPC = ____________
3. EA = 12 m; Area of sector BPA = ____________
4. PG = 10 m; Area of sector GPA = ____________
5. FP = 16 m; Area of sector GPF = ____________
6. EA = 36 m; Area of sector CPB = ____________

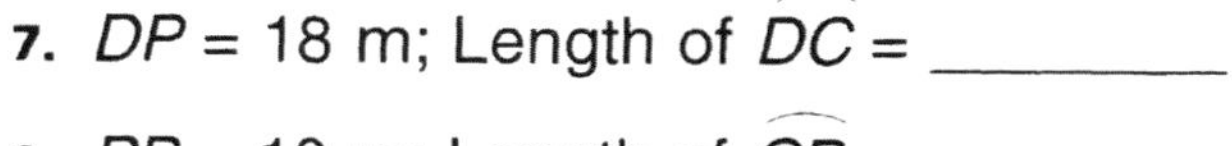

7. DP = 18 m; Length of $\overset{\frown}{DC}$ = __________
8. PB = 10 m; Length of $\overset{\frown}{CB}$ = __________
9. EA = 24 m; Length of $\overset{\frown}{FA}$ = __________
10. GP = 15 m; Length of $\overset{\frown}{FG}$ = __________
11. EA = 16 m; Length of $\overset{\frown}{EG}$ = __________
12. PD = 30 m; Length of $\overset{\frown}{DA}$ = __________

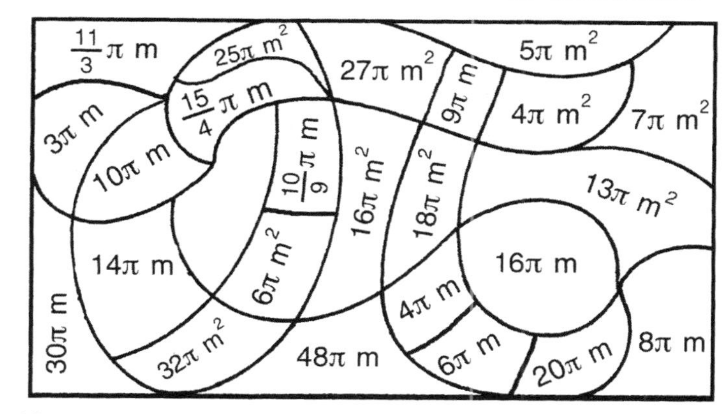

© Milliken Publishing Company

Name ______________________________

Surface Area and Volume of Spheres

Remember

A **sphere** is a solid. All the points on the surface of a sphere are an equal distance from the center (C) of the sphere. That distance is the radius (r).

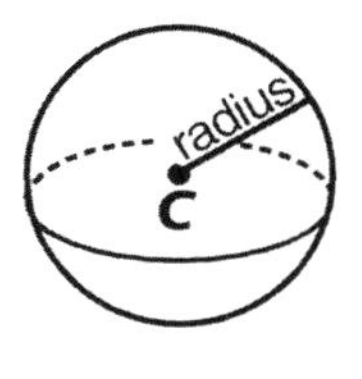

Surface Area of a sphere = $4\pi r^2$ square units

Volume of a sphere = $\frac{4}{3}\pi r^3$ cubic units

$SA = 4\pi r^2$ $\qquad$ $V = \frac{4}{3}\pi r^3$

Example:
Find the surface area and volume of this sphere.

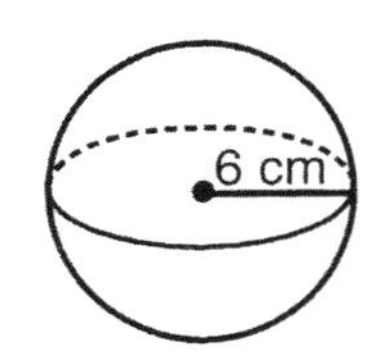

$$SA = 4\pi r^2 = 4 \cdot \pi \cdot 6^2 = 4 \cdot 36 \cdot \pi = 144\pi \text{ cm}^2$$

$$V = \frac{4}{3}\pi r^3 = \frac{4}{3} \cdot \pi \cdot 6^3 = \frac{4}{3}(216)\pi = 288\pi \text{ cm}^3$$

Draw straight lines to match each radius of a sphere to its correct surface area and volume. Write the uncrossed letters in the empty circles below to reveal the name of the mathematician who made the first scientific attempt to compute the value of π.

Surface Area	Radius	Volume
400π cm^2	6 cm	2304π in.^{3}
36π ft^2	15 mm	4500π mm^3
900π mm^2	10 cm	$\frac{16,384}{3}\pi$ in.^{3}
144π cm^2	12 in.	$\frac{4000}{3}\pi$ cm^3
1296π cm^2	3 ft	288π cm^3
324π mm^2	16 in.	7776π cm^3
576π in.^{2}	25 ft	972π mm^3
2500π ft^2	9 mm	36π ft^3
1024π in.^{2}	18 cm	$\frac{62,500}{3}\pi$ ft^3

Letters between Surface Area and Radius: A, U, L, R, C, H, D, K, E, I

Letters between Radius and Volume: M, J, A, E, T, D, E, N, O, S

Name ______________________________

Date ________________ Score _______ %

Assessment A

Geometry Basics

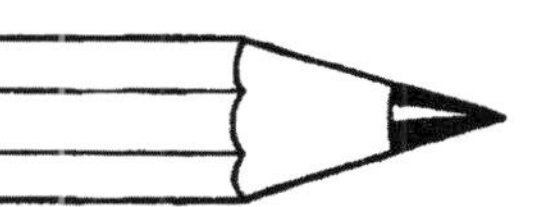

Shade in the circle of the correct answer.

1. Which two lines are coplanar?

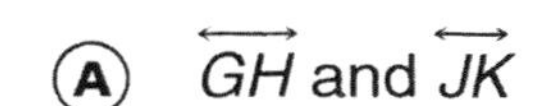

Ⓐ $\overleftrightarrow{GH}$ and $\overleftrightarrow{JK}$

Ⓑ $\overleftrightarrow{JK}$ and $\overleftrightarrow{MN}$

Ⓒ $\overrightarrow{LK}$ and $\overrightarrow{LJ}$

Ⓓ J and K

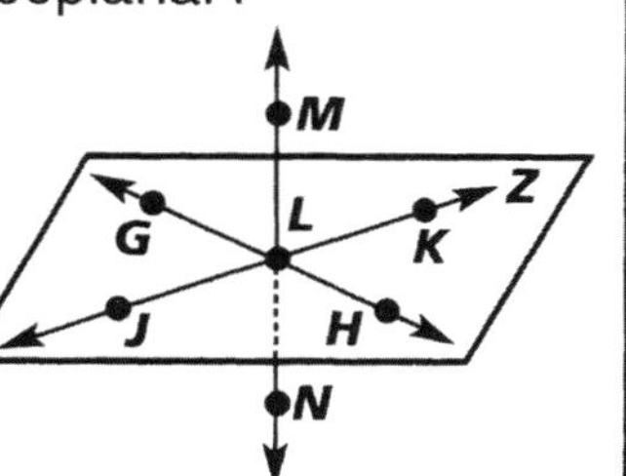

2. Which point is the vertex of $\angle HLK$?

Ⓐ G Ⓑ H

Ⓒ L Ⓓ K

3. Which describes these two angles?

Ⓐ complementary

Ⓑ supplementary

Ⓒ vertical

Ⓓ None of these answers.

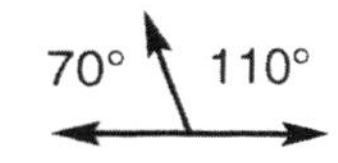

4. These two angles are complementary. What is the measure of x?

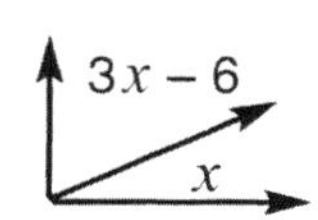

Ⓐ 24° Ⓑ 32°

Ⓒ 45° Ⓓ 90°

5. Which pair of angles is congruent?

Ⓐ $\angle 1$ and $\angle 2$

Ⓑ $\angle 1$ and $\angle 7$

Ⓒ $\angle 5$ and $\angle 7$

Ⓓ $\angle 5$ and $\angle 8$

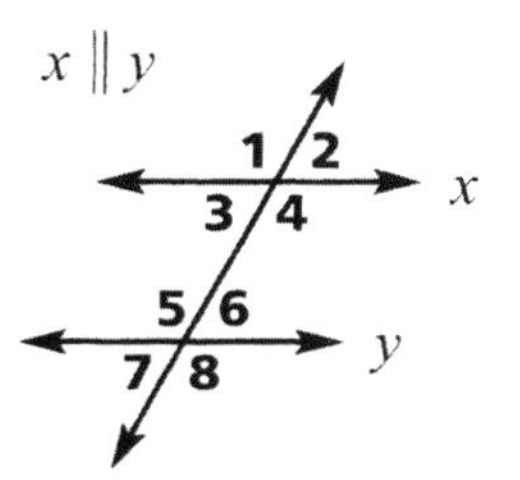

6. Given: $\overrightarrow{RT}$ bisects $\angle QRS$. $m\angle TRS = 25°$. Which is true?

Ⓐ $m\angle QRS = 25°$

Ⓑ $m\angle QRT = 50°$

Ⓒ $m\angle QRS = 50°$

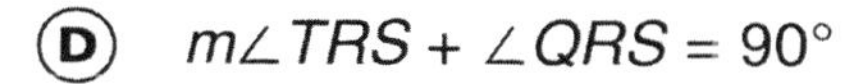

Ⓓ $m\angle TRS + \angle QRS = 90°$

7. Given: $\overleftrightarrow{GK}$ bisects $\overline{EF}$. $\overrightarrow{JL}$ bisects $\overline{HG}$. $EF = 20$. $HJ = 3$. How long is $\overline{EH}$?

Ⓐ 3

Ⓑ 4

Ⓒ 5

Ⓓ 10

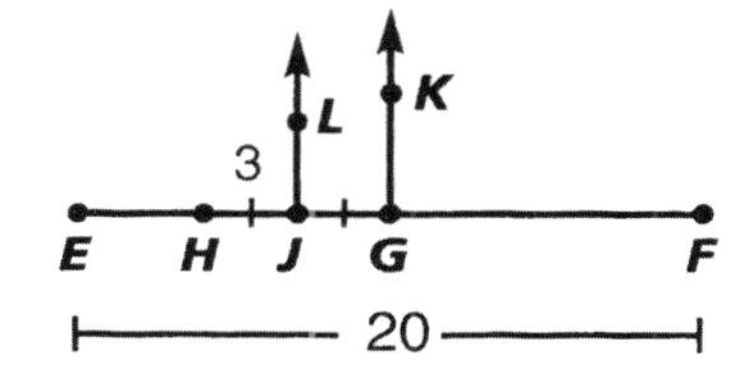

8. Which triangle has these vertices: (1, ⁻1), (1, ⁻4), and (3, ⁻4)?

Ⓐ $\triangle ABC$

Ⓑ $\triangle DEF$

Ⓒ $\triangle JKL$

Ⓓ $\triangle RST$

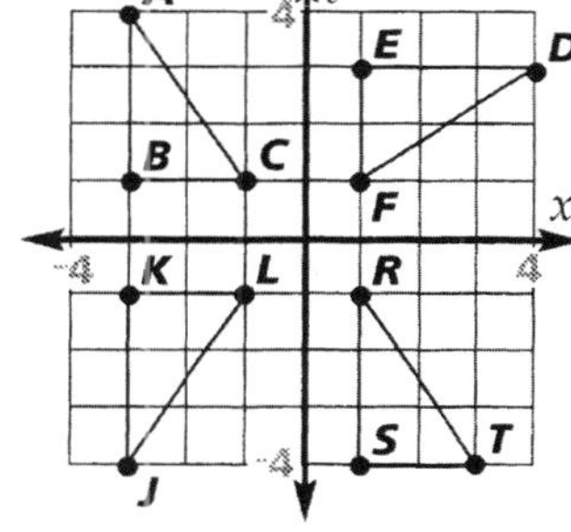

9. Which triangle is a reflection of $\triangle$**ABC** over the x-axis?

Ⓐ $\triangle ABC$ Ⓑ $\triangle DEF$

Ⓒ $\triangle JKL$ Ⓓ $\triangle RST$

10. Which point is the midpoint between (⁻1, 1) and (⁻5, ⁻7)?

Ⓐ (⁻6, ⁻8) Ⓑ (⁻6, ⁻6)

Ⓒ (⁻4, ⁻6) Ⓓ (⁻3, ⁻3)

© Milliken Publishing Company

Name ____________________________

Date ________________ Score ______ %

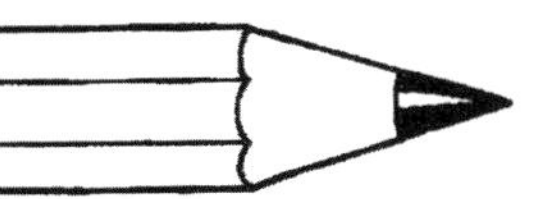

Assessment B
Triangles

Shade in the circle of the correct answer.

1. What is the measure of $\angle Z$?

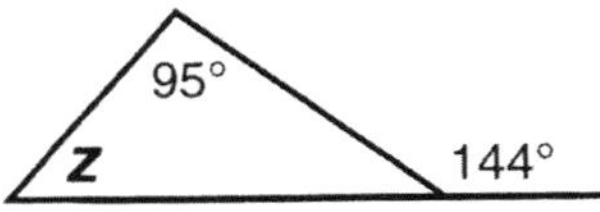

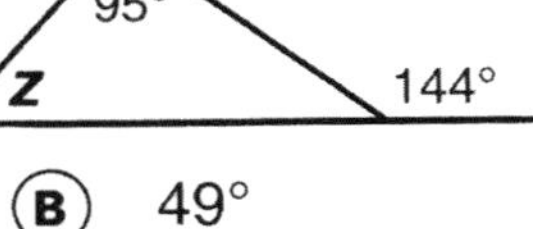

Ⓐ 36° Ⓑ 49°

Ⓒ 131° Ⓓ 144°

2. Which side is shortest?

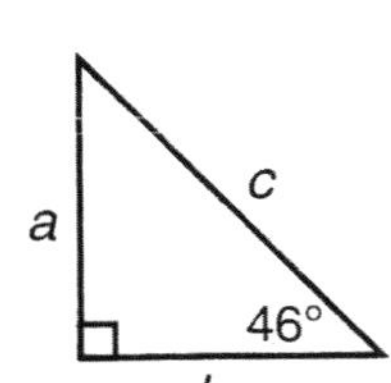

Ⓐ *a*

Ⓑ *b*

Ⓒ *c*

Ⓓ cannot be determined

3. Which correctly describes this triangle?

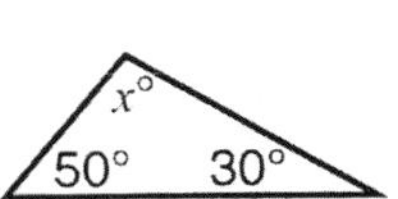

Ⓐ scalene and obtuse

Ⓑ isosceles and acute

Ⓒ scalene and acute

Ⓓ right and equilateral

4. Which segment is an altitude of $\triangle ABC$?

E, F, and ***G*** are midpoints of $\overline{AB}$, $\overline{AC}$, and $\overline{BC}$.

Ⓐ $\overline{AC}$

Ⓑ $\overline{AG}$

Ⓒ $\overline{EF}$

Ⓓ None of these answers.

A, E, F, B, D, G, C

5. If $EF = 10$, which is true?

Ⓐ $AD = 10$ Ⓑ $BC = 10$

Ⓒ $BC = 20$ Ⓓ $EG = GF$

6. Which congruence method(s) could be used to prove the triangles are congruent?

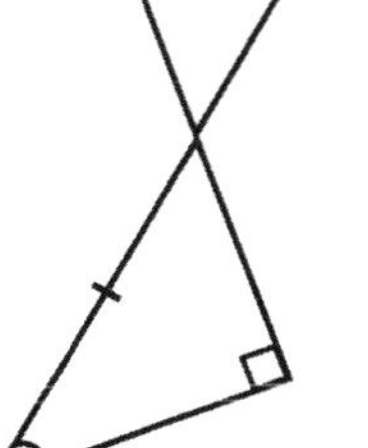

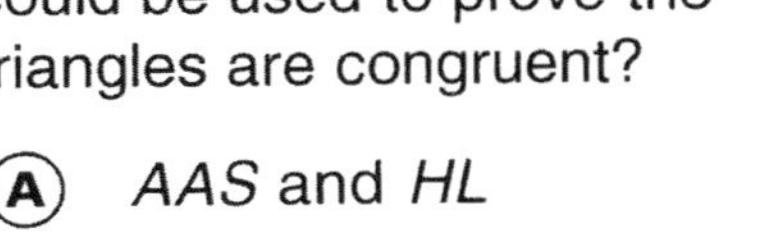

Ⓐ *AAS* and *HL*

Ⓑ *HL* only

Ⓒ *ASA* only

Ⓓ *AAS* and *ASA*

7. If $c^2 = a^2 + b^2$, what type of triangle is this?

a, b, c

Ⓐ right Ⓑ acute

Ⓒ obtuse Ⓓ equilateral

8. What is the length of the missing side?

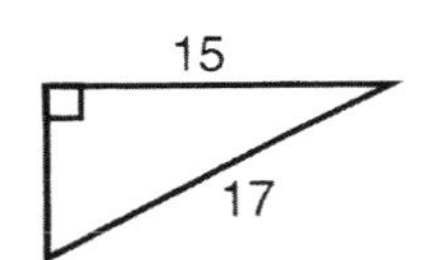

Ⓐ 8 Ⓑ 9

Ⓒ 11 Ⓓ 13

9. What is the distance between these two points: (⁻3, 10) and (1, 7)?

Ⓐ $\sqrt{7}$ units Ⓑ 4 units

Ⓒ 5 units Ⓓ 25 units

10. What is the measure of side *g*?

Ⓐ $3\sqrt{2}$

Ⓑ $3\sqrt{3}$

Ⓒ 6

Ⓓ cannot be determined

3, 60°, 30°, g

© Milliken Publishing Company

Name ______________________________

Date ________________ Score ______ %

Assessment C

Trigonometry, Polygons, and Solids

Shade in the circle of the correct answer.

1. Which trigonometric equation is correct?

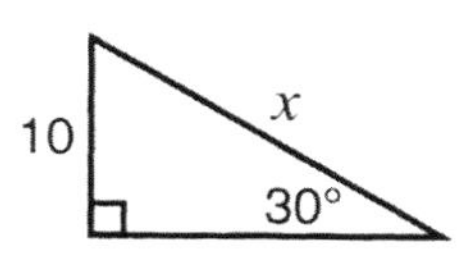

Ⓐ $\sin 30° = \frac{10}{x}$ Ⓑ $\cos 30° = \frac{10}{x}$

Ⓒ $\tan 30° = \frac{x}{10}$ Ⓓ $\sin 60° = \frac{x}{10}$

2. Use the formula to determine the measure of an interior angle in a regular hexagon.

$$\frac{(n-2) \cdot 180°}{n}$$

Ⓐ 30°

Ⓑ 60°

Ⓒ 110°

Ⓓ 120°

3. This shape is both a rectangle and a

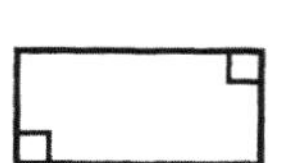

Ⓐ parallelogram. Ⓑ rhombus.

Ⓒ trapezoid. Ⓓ square.

4. What is the length of the long base?

Ⓐ 8

Ⓑ 10

Ⓒ 16

Ⓓ 22

5. These figures are similar. What is the length of z?

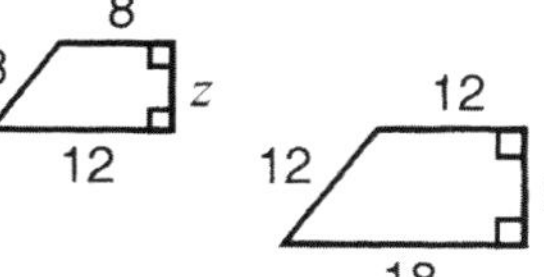

Ⓐ 3 Ⓑ 4

Ⓒ 6 Ⓓ 8

6. What is the perimeter of this figure?

Ⓐ 32 cm

Ⓑ 35 cm

Ⓒ 64 cm

Ⓓ 70 cm

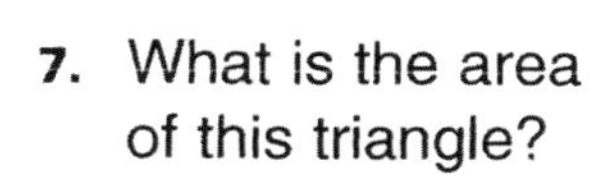

7. What is the area of this triangle?

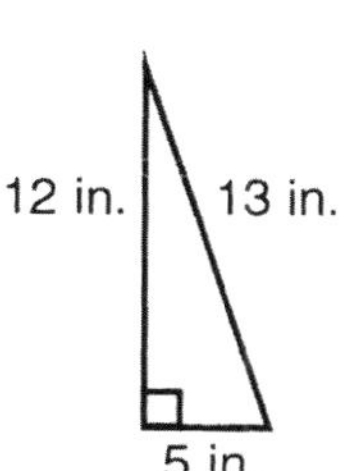

Ⓐ 30 in.^{2} Ⓑ 32 in.^{2}

Ⓒ 60 in.^{2} Ⓓ 65 in.^{2}

8. Use the formula to determine the area of this trapezoid.

$$A = \frac{1}{2}h(b_1 + b_2)$$

Ⓐ 24 cm^2

Ⓑ 30 cm^2

Ⓒ 42 cm^2

Ⓓ 84 cm^2

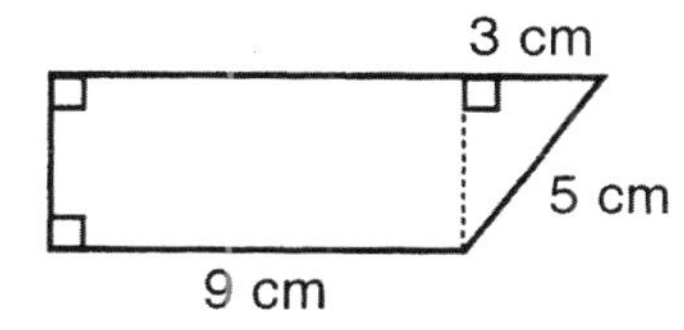

9. What is the surface area of this right prism?

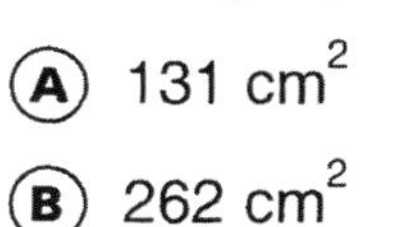

Ⓐ 131 cm^2

Ⓑ 262 cm^2

Ⓒ 280 cm^2

Ⓓ 560 cm^2

10. Which volume formula should be used for a right triangular prism?

Ⓐ $V = s^3$ Ⓑ $V = Bh$

Ⓒ $V = lwh$ Ⓓ $V = \frac{4}{3}\pi r^3$

© Milliken Publishing Company

Name ______________________________

Date ________________ Score ______ %

Assessment D

Circles

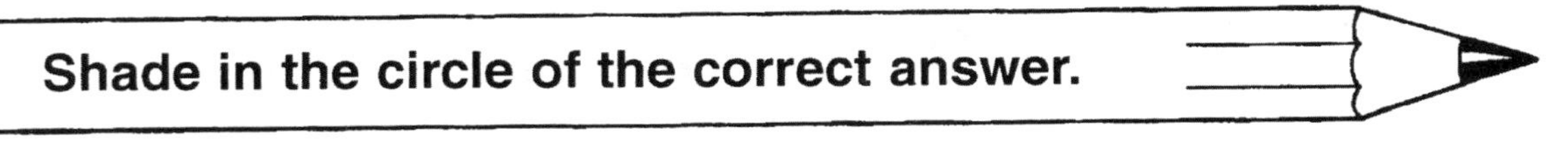

1. What is the circumference of this circle? Use 3.14 for π.

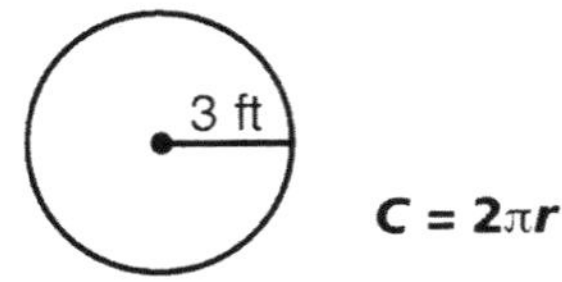

$C = 2\pi r$

Ⓐ 9 ft Ⓑ 10 ft

Ⓒ 18 ft Ⓓ 19 ft

2. What is the area of the shaded region?

Area of a circle = πr^2

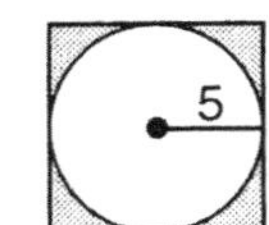

Ⓐ $(25 - 25\pi)$ units^2

Ⓑ $(100 - 25\pi)$ units^2

Ⓒ $(5\pi - 10)$ units^2

Ⓓ $(25\pi + 100)$ units^2

3. What is the volume of this cylinder?

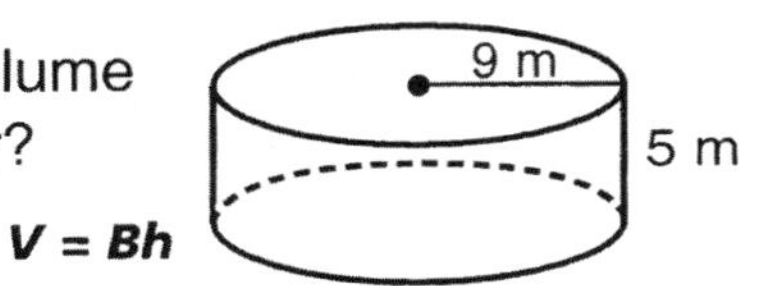

$V = Bh$

Ⓐ 45 m^2 Ⓑ 45π m^2

Ⓒ 81π m^3 Ⓓ 405π m^3

4. Given: C is the center. $\overline{XZ}$ is a diameter. Which arc measures 245°?

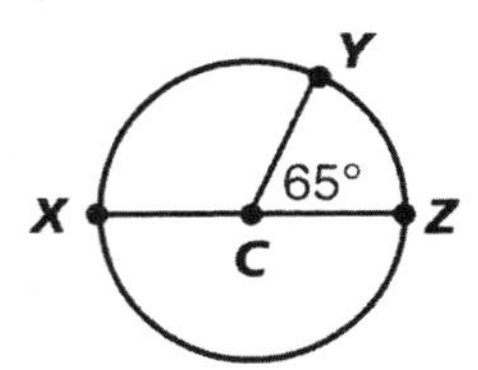

Ⓐ $\overparen{XY}$ Ⓑ $\overparen{YZ}$

Ⓒ $\overparen{YZX}$ Ⓓ $\overparen{ZXY}$

5. What is the measure of $\angle PRQ$?

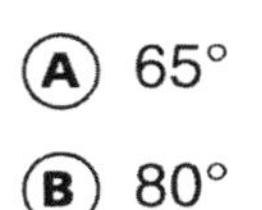

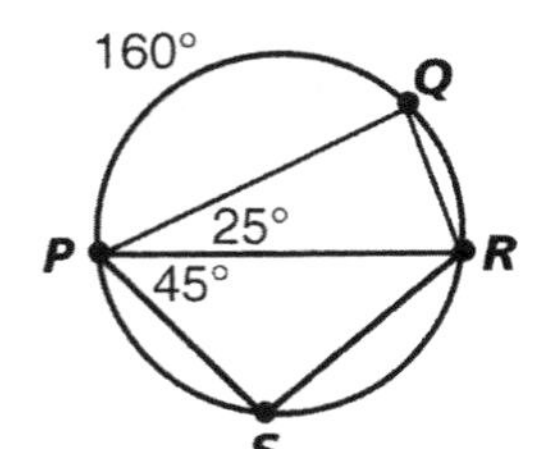

Ⓐ 65°

Ⓑ 80°

Ⓒ 155°

Ⓓ 160°

6. This diagram shows the intersection of two

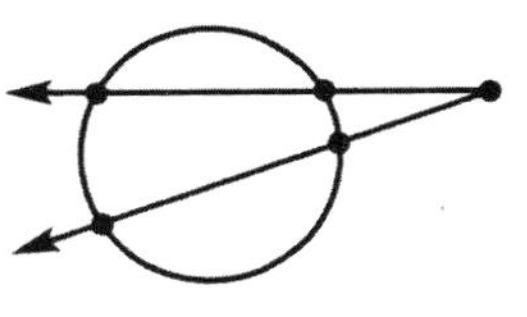

Ⓐ circles. Ⓑ chords.

Ⓒ secants. Ⓓ tangents.

7. Given: $m\overparen{RS} = 135°$. What is the measure of $\angle 1$?

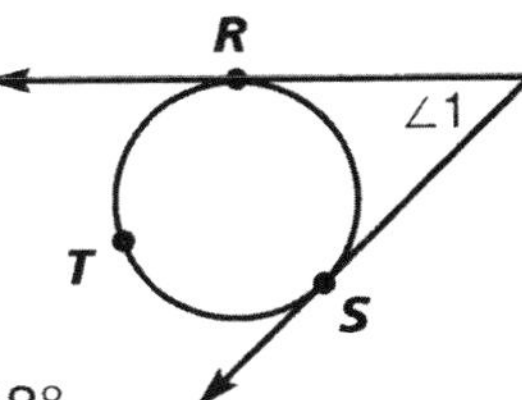

Ⓐ 45° Ⓑ 48°

Ⓒ 67.5° Ⓓ 225°

8. What is the length of $\overline{EG}$?

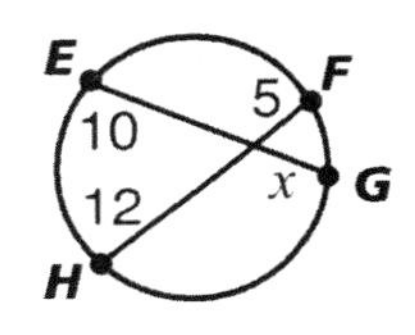

Ⓐ 6 Ⓑ 7

Ⓒ 16 Ⓓ 17

9. What is the length of $\overline{NM}$?

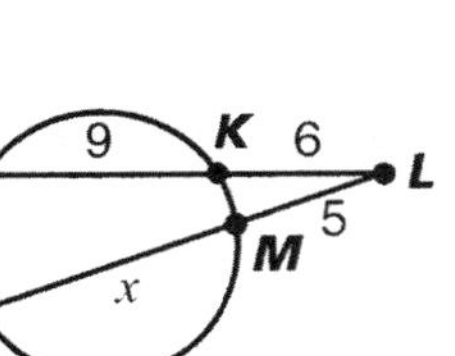

Ⓐ 13

Ⓑ 14

Ⓒ 15

Ⓓ 16

10. Given: C is the center. What is the area of the shaded sector?

120°
C
10 cm

$A = \frac{m}{360°} \cdot \pi r^2$

Ⓐ $\frac{1}{3}\pi$ cm^2

Ⓑ $\frac{10}{3}\pi$ cm^2

Ⓒ 100π cm^2

Ⓓ None of these answers.

Answers

Page 3

1. true
2. false
3. true
4. true
5. false
6. true
7. false
8. false
9. true
10. true
11. true
12. true
13. true
14. false
15. true
16. true
17. false
18. true
19. false
20. true
21. true
22. true

EUCLID'S ELEMENTS

Page 4

1. right
2. acute
3. straight
4. obtuse
5. acute
6. right
7. straight
8. obtuse
9. complementary
10. supplementary
11. complementary
12. neither
13. neither
14. supplementary
15. neither
16. supplementary

Angles that have the same measure are called congruent angles.

Page 5

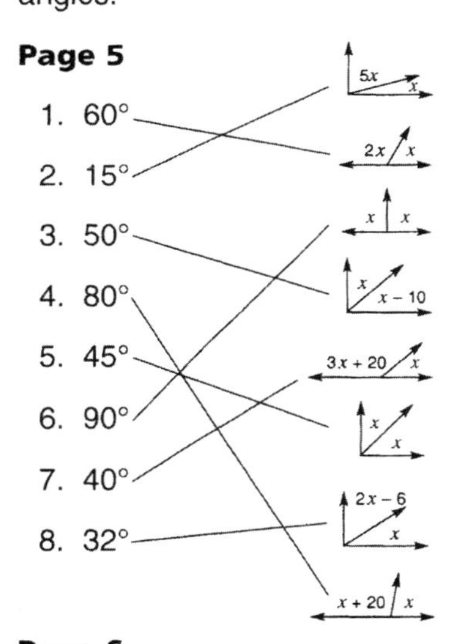

Page 6

1. 60° adjacent
2. 130° vertical
3. 55° interior
4. 125° corresponding
5. 45° exterior
6. 50° alternate
7. 138° same side
8. 90° right

Page 7

1. $m\angle ABD = 28°$ $m\angle ABC = 56°$
2. $m\angle EFH = 55°$ $m\angle HFG = 55°$
3. $m\angle IJL = 35°$ $m\angle IJK = 70°$
4. $m\angle PNO = 50°$ $m\angle MNO = 100°$
5. $m\angle TRS = 42°$ $m\angle QRS = 84°$
6. $EC = 4$ $AB = 22$ $AD = 3$
7. $UW = 20$ $YV = 40$ $UV = 80$

MIDPOINT

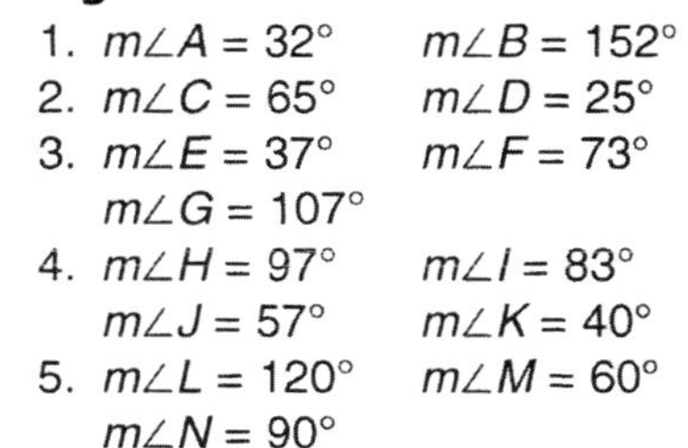

Page 8

1. $m\angle A = 32°$ $m\angle B = 152°$
2. $m\angle C = 65°$ $m\angle D = 25°$
3. $m\angle E = 37°$ $m\angle F = 73°$
   $m\angle G = 107°$
4. $m\angle H = 97°$ $m\angle I = 83°$
   $m\angle J = 57°$ $m\angle K = 40°$
5. $m\angle L = 120°$ $m\angle M = 60°$
   $m\angle N = 90°$
6. $m\angle O = 79°$ $m\angle P = 12°$

Page 9

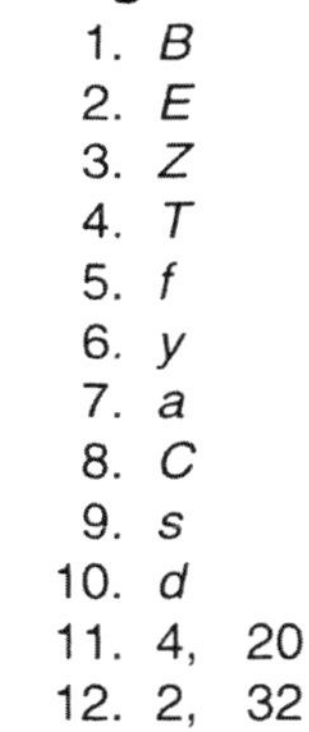

1. *B*
2. *E*
3. *Z*
4. *T*
5. *f*
6. *y*
7. *a*
8. *C*
9. *s*
10. *d*
11. 4, 20
12. 2, 32
13. 5, 45
14. 1, 7
15. 0, 18

Page 10

1. 90°; right, scalene
2. 8; obtuse, isosceles
3. 60°; equiangular, acute, equilateral, isosceles
4. 70°; acute, scalene
5. 75°; acute, isosceles
6. 10; right, isosceles
7. 110°; obtuse, scalene
8. 70°; 15; acute, isosceles

PYTHAGOREAN THEOREM

Page 11

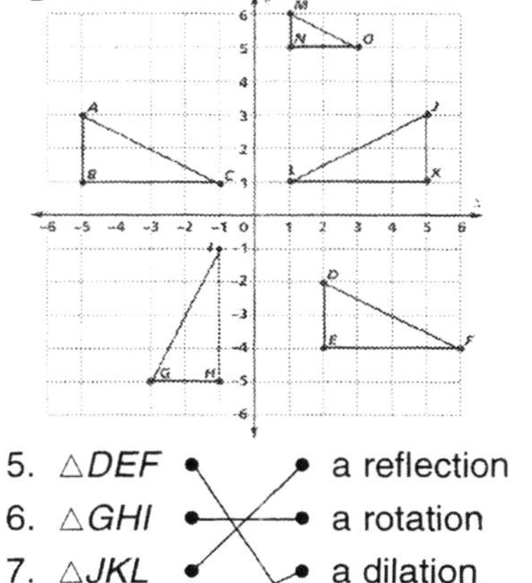

5. $\triangle DEF$ — a reflection
6. $\triangle GHI$ — a rotation
7. $\triangle JKL$ — a dilation
8. $\triangle MNO$ — a translation

SUPER!

Page 12

1. (10, 6)
2. (0, 3)
3. (⁻6.5, ⁻4)
4. (5, 7)
5. (⁻3, ⁻3)
6. (⁻6, ⁻1)
7. (5.5, 5)

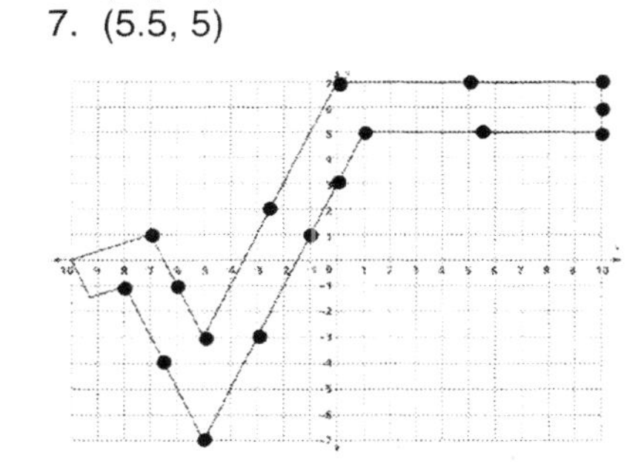

Page 13

1. altitude
2. median
3. midline
4. *CE*
5. *BC*
6. 8
7. 6
8. *JK*
9. 10
10. 15
11. *JL*
12. 30

The Babylonians

Page 14

1. SSS
2. SAS
3. ASA
4. ASA
5. SSS
6. SAS
7. can't
8. ASA
9. can't
10. SAS
11. can't
12. SSS

PERFECT PAPER!

Page 15

1. HL
2. AAS
3. SAS, ASA, AAS
4. AAS, HL
5. SSS, SAS
6. HL
7. ASA, AAS
8. can't
9. SAS, ASA, AAS, HL

NONE. THEY CAN'T DO IT. THEY CAN ONLY PROVE IT CAN BE DONE!

Page 16

1. 1. • • SAS Congruence
   2. • • SSS Congruence
   3. • • Given
   4. • • Given
   • Reflexive Property

2. 1. • • Reflexive Property
   2. • • AAS Congruence
   3. • • Alt. Interior Angles
   4. • • SAS Congruence
   5. • • Rt. ∠ Congruence
   • Given

3. 1. • • Vertical Angles
   2. • • SAS Congruence
   3. • • Given
   4. • • Given
   • SSS Congruence

4. 1. • • Alt. Interior Angles
   2. • • AAS Congruence
   3. • • Reflexive Property
   4. • • Given
   • ASA Congruence

5. 1. • • Def. of Rt. Triangles
   2. • • Given
   • SAS Congruence
   3. • • HL Congruence
   4. • • Reflexive Property
   5. • • Given

AWESOME!

Page 17

A. 3
E. 15
G. $\sqrt{13}$
H. 8
I. $4\sqrt{2}$
L. 12
N. 10
O. 25
P. 20
R. 4
S. $3\sqrt{3}$
T. 13
Y. 26

PYTHAGOREAN TRIPLES

Page 18

1. right
2. no triangle
3. acute
4. obtuse
5. acute
6. obtuse
7. right
8. acute
9. right
10. obtuse
11. right
12. no triangle
13. obtuse
14. acute

JAMES A. GARFIELD

Page 19

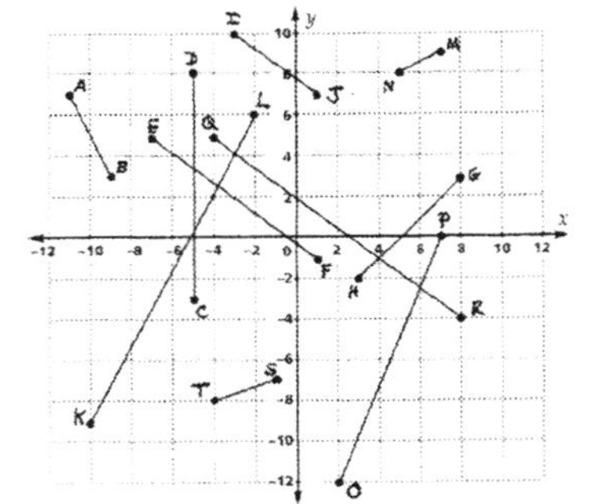

1. $2\sqrt{5}$
2. 11
3. 10
4. $5\sqrt{2}$
5. 5
6. 17
7. $\sqrt{5}$
8. 13
9. 15
10. $\sqrt{10}$

SIXTY FEET, SIX INCHES

Page 20

1. $8; 8\sqrt{2}$
2. $5\sqrt{3}; 10$
3. $6\sqrt{3}; 6$
4. $3; 3$
5. $4; 2$
6. $\sqrt{6}; \sqrt{3}$
7. $\sqrt{2}; \sqrt{2}$
8. $\sqrt{15}; 2\sqrt{5}$
9. $\frac{8\sqrt{3}}{3}; \frac{16\sqrt{3}}{3}$
10. $2\sqrt{3}; \sqrt{3}$

Page 21

1. sin *B* • • $\frac{\text{adj.}}{\text{hyp.}}$
   cos *B* • • $\frac{\text{opp.}}{\text{adj.}}$
   tan *B* • • $\frac{\text{opp.}}{\text{hyp.}}$

2. sin 30° • • $\frac{\sqrt{3}}{2}$
   sin 60° • • $\frac{\sqrt{3}}{1}$
   tan 30° • • $\frac{1}{\sqrt{3}}$
   tan 60° • • $\frac{1}{2}$

3. cos 45° • • $\frac{\text{opp.}}{\text{hyp.}}$
   tan 45° • • $\frac{\text{adj.}}{\text{hyp.}}$
   sin 45° • • $\frac{1}{1}$

4. tan *D* • • $\frac{12}{13}$
   tan *E* • • $\frac{12}{5}$
   cos *D* • • $\frac{5}{12}$

YOU'RE A TRIG EXPERT!

Page 22

1. $\cos 25° = \frac{60}{x}$; 66 ft
2. $\tan 30° = \frac{x}{150}$; 87 ft
3. $\sin 15° = \frac{48}{x}$; 185 ft
4. $\tan 50° = \frac{6}{x}$; 5 ft
5. $\sin 65° = \frac{x}{72}$; 65 ft

© Milliken Publishing Company

Page 23

t	3	1	180°	60°	360°	120°
q	4	2	360°	90°	360°	90°
p	5	3	540°	108°	360°	72°
h	6	4	720°	120°	360°	60°
o	8	6	1080°	135°	360°	45°
de	10	8	1440°	144°	360°	36°
do	12	10	1800°	150°	360°	30°

Page 24

1. a. 14 c. 52° e. 32°
   b. 11 d. 128° f. 88°
2. a. 4 c. 60° e. $2\sqrt{3}$
   b. 30° d. 2 f. 120°
3. a. 90° c. 37° e. 106°
   b. 20 d. 53° f. 10
4. a. 8 c. 45° e. $4\sqrt{2}$
   b. 90° d. $8\sqrt{2}$ f. 45°

■ (SQUARE) ROOTS

Page 25

1. $a = 8$, $b = 5$, $c = 5\sqrt{3}$, $d = 13$
2. $e = 3\sqrt{3}$, $f = 3$, $g = 12$, $h = 18$
3. $j = 2\sqrt{3}$, $k = 2\sqrt{6}$, $m = 2\sqrt{3}$, $n = 10$, $p = 4$
4. $q = 15$, $r = 17$, $s = 7$
5. $t = 16$, $u = 20$, $v = 12$, $w = 44$
6. $x = 4$, $y = 14$, $z = 4\sqrt{2}$

NEVADA

Page 26

1. $F = 9$; $G = 21$
2. $L = 7$; $M = 6$
3. $D = 15$; $E = 20$
4. $B = 10$
5. $R = 12.5$; $T = 16$
6. $W = 18$
7. $H = 36$; $I = 5$
8. $Z = 12$
9. $N = 13.5$; $O = 22.5$

GOTTFRIED WILHELM LEIBNIZ

Page 27

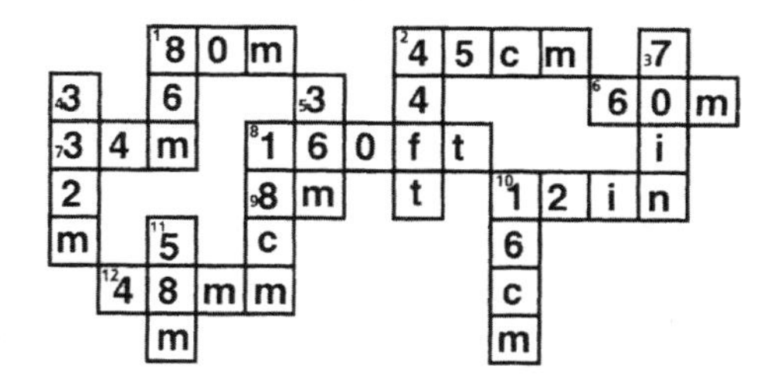

Page 28

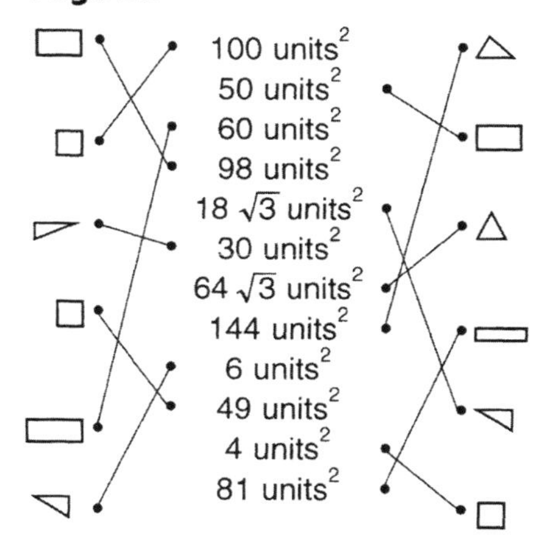

BRAVO!

Page 29

1. 240 cm^2
2. 46 ft^2
3. 138 m^2
4. 8 $in.^2$
5. 40 m^2
6. 72 $in.^2$
7. 36 cm^2
8. 144 m^2
9. 165 ft^2
10. 42 cm^2
11. 180 ft^2
12. 20 $in.^2$

Page 30

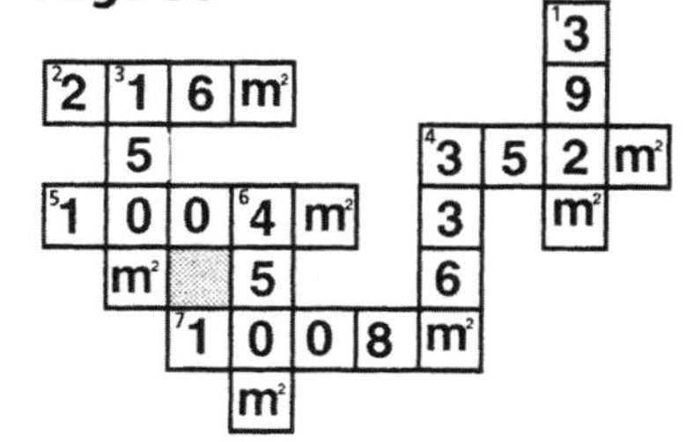

Page 31

1. 360 cm^3
2. 240 m^3
3. 2,100 m^3
4. 729 cm^3
5. 540 cm^3
6. 280 cm^3
7. 1,440 m^3

OMAR KHAYYAM

Page 32

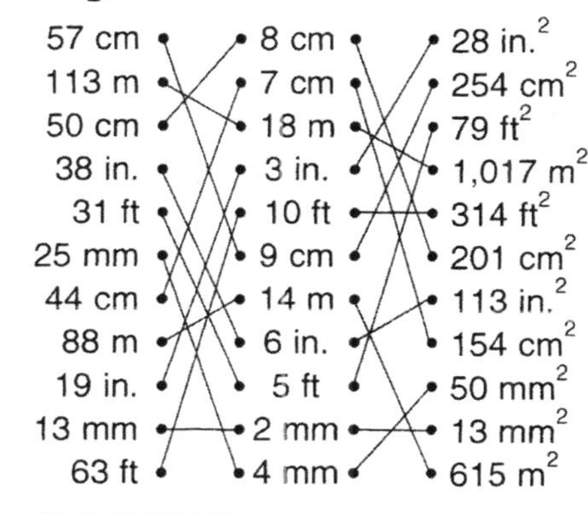

PUMPKIN PI

Page 33

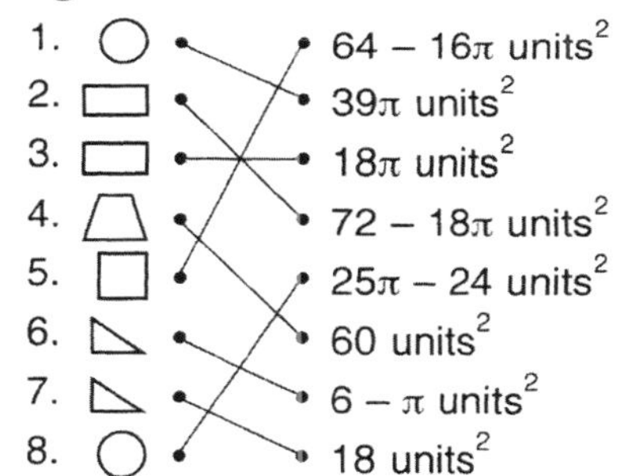

Page 34

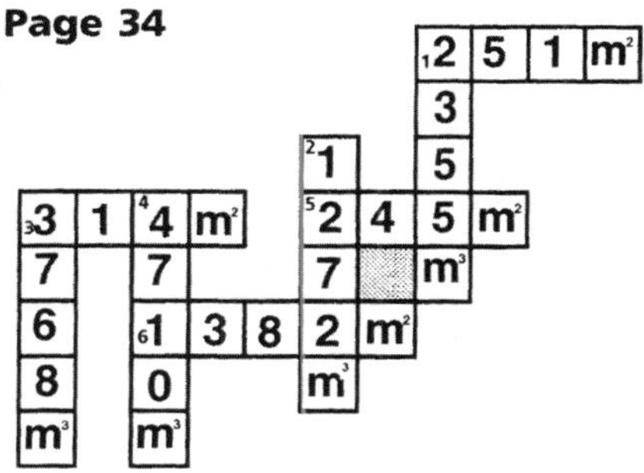

Page 35

1. $m\overset{\frown}{XY}$	180°
2. $m\angle XOZ$	310°
3. $m\angle YOZ$	50°
4. $m\overset{\frown}{XZY}$	230°
5. $m\overset{\frown}{YXZ}$	130°
6. $m\overset{\frown}{JL}$	90°
7. $m\angle KLJ$	95°
8. $m\angle KJL$	85°
9. $m\overset{\frown}{LM}$	60°
10. $m\angle KML$	65°
11. $m\overset{\frown}{BC}$	100°
12. $m\angle BEC$	80°
13. $m\overset{\frown}{BEC}$	40°
14. $m\overset{\frown}{DC}$	20°
15. $m\angle EBD$	280°

Page 36

1. 75°
2. 55°
3. 40°
4. 25°
5. 30°
6. 115°
7. 50°
8. 90°
9. 80°
10. 95°
11. 65°
12. 70°

BECAUSE HE WANTED TO BE A TAN GENT!

Page 37

1. 10; 12
2. 6; 17
3. 9; 22
4. 7; 7.5
5. 10.5; 14.5
6. 16; 19
7. 6; 13
8. 4; 10
9. 12; 24
10. 2; 9
11. 3; 18
12. 5; 15

Page 38

1. 22; 25; 10
2. 24; 4; 12
3. 40; 20
4. 4.4; 14.4
5. 16; 9
6. 17; 7

MARIA AGNESI;
WITCH OF AGNESI

Page 39

1. 6π m^2
2. 27π m^2
3. 4π m^2
4. 25π m^2
5. 32π m^2
6. 18π m^2
7. 6π m
8. $\frac{10}{9}\pi$ m
9. 9π m
10. $\frac{15}{4}\pi$ m
11. 4π m
12. 20π m

Page 40

ARCHIMEDES

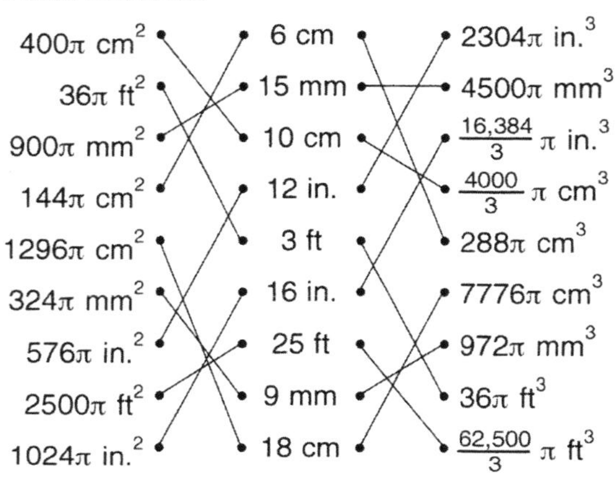

Page 41

1. A
2. C
3. B
4. A
5. D
6. C
7. B
8. D
9. C
10. D

Page 42

1. B
2. B
3. A
4. D
5. C
6. D
7. A
8. A
9. C
10. B

Page 43

1. A
2. D
3. A
4. D
5. C
6. D
7. A
8. C
9. B
10. B

Page 44

1. D
2. B
3. D
4. C
5. B
6. C
7. A
8. C
9. A
10. D

© Milliken Publishing Company